BRAIN APPS

HACKING NEUROSCIENCE TO *GET THERE*

Robert G. Best

with J.M. Best

To
Rich

Best Wishes

Parts of this book appeared in different form on Robert G. Best's blog site, *Mindframewithrobb*

Audio book link available at *BestMindframe.com*, a division of Giant Pivot Table LLC.

Cover & book design by Vincent Skyers of UmbraSolutions.com

Edited by J.M. Best

Published by BookLocker.com, Inc., St. Petersburg, Florida, U.S.A.

Printed on acid-free paper.

BookLocker.com, Inc.

2017

First Edition

DISCLAIMER

This book details the author's personal experiences with and opinions about cognitive science. The author is not a licensed medical doctor.

The author and publisher are providing this book and its contents on an "as is" basis and make no representations or warranties of any kind with respect to this book or its contents. The author and publisher disclaim all such representations and warranties, including for example warranties of merchantability and specific advice for a particular purpose. In addition, the author and publisher do not represent or warrant that the information accessible via this book is accurate, complete or current.

The statements made about products and services have not been evaluated by the U.S. government. Please consult with your own legal, accounting, medical, or other licensed professional regarding the suggestions and recommendations made in this book.

Except as specifically stated in this book, neither the author or publisher, nor any authors, contributors, or other representatives will be liable for damages arising out of or in connection with the use of this book. This is a comprehensive limitation of liability that applies to all damages of any kind, including (without limitation) compensatory; direct, indirect or consequential damages; loss of data, income or profit; loss of or damage to property and claims of third parties.

You understand that this book is not intended as a substitute for consultation with a licensed medical, legal or accounting professional. Before you begin any change your lifestyle in any way, you will consult a licensed professional to ensure that you are doing what's best for your situation.

This book provides content related to cognitive science topics. As such, use of this book implies your acceptance of this disclaimer.

Contents

Prologue

What separates the supremely accomplished from the rest of us? Is genius learned or innate? Those questions have haunted us for decades. About four years ago, we embarked on a mission to find answers. The more we delved into the mysteries of what makes people successful and how—or even if—those traits can be learned, the more two things quickly became apparent.

First, that it all seemed to lead back to neuroscience. To improve your behavior, skills, and thinking—to truly pursue excellence—you must maximize your brain's potential. Learning how to do that means both looking at some surprising new insights about how the brain works, and unlearning some harmful myths.

Second, we discovered that these lessons are startlingly universal. When we studied people operating at the top of their game, whether it was business, music, sports, or science, a few common truths seemed to emerge.

You can't always choose your circumstances or experience the bright lights of fame. But if your goal is to radically improve your skills and master something successfully, that is absolutely within your control. You can create brain apps that will take you wherever you want to go.

This book can be read in two ways. You can approach it as a curious observer, pulling back the curtain on genius and the path people have followed to get there. Or, you can read it as a how-to, a blueprint for the steps necessary to harness your own brainpower for your best chance of success at work, play or life.

Either way, you'll likely pick up a few brain hacks you'll find useful.

We hope you enjoy reading this book as much as we enjoyed researching and writing it.

Robert G. Best

J.M. Best

Chapter 1
Mindset App

Imagine that we can travel back in time to 1995. We're in New York City, standing on the steps of Manhattan's Fiorello H. LaGuardia High School.

If La Guardia High sounds familiar, you might recognize it as the School of the Performing Arts from the movie Fame. That said, there's nothing glamorous about its appearance, a giant boxy structure of cement and glass. The building is old and beat-up; it feels more like a prison than a high school. You walk through the double doors and after a couple of twists and turns, you climb down a flight of concrete stairs into the basement. From here, we make our way toward the back of the building, where you can hear the school band practicing.

How do they sound? Like every high school band in every state, in every city or town of this country: mixed. That is to say, the players run the full spectrum: some of them are really good, and some are only average.

And within four years, the vast majority of them will have moved on to something else. The instruments they're playing now, in 1995, will wind up in attics, garages and crawl spaces. Not a lot of kids keep pursuing the tuba or the piccolo after they've landed their first full-time job. Not a lot of kids take their timpani to college. The horns and drums are more likely to be sold, thrown out or, for the more desperate, pawned.

But at least one of those instruments in La Guardia High School will not end up in a pawnshop. It belongs to a guy in the brass section—not just any guy, but Dave Guy.

It's a bit of fluke that Dave even got this far. He'd wanted to play trumpet since seventh grade, but the position was already filled, and his junior high band teacher talked him into playing the clarinet instead.

CHAPTER 1

And so Dave began his musical career as a very half-hearted clarinetist. He wasn't interested, rarely practiced, and as a result, never got that good. Our story could end right here, with Dave Guy as just another former band kid, except for one lucky thing: in the middle of the year, the band director left unexpectedly.

When the new band teacher showed up, Dave didn't mess around.

"Nice to meet you," he said. "I'm Dave Guy, I'm the trumpet player."

Why was young Dave so determined to follow this path? Had the legendary Louis Armstrong inspired him? Was he enchanted by the music of Miles Davis or Chet Baker?

Nope.

When we interviewed him later, Dave gave three reasons for his early interest in the trumpet:

First, it was shiny. Second, he thought it had a cool-looking shape. Last, and most importantly, "It was really loud!"

Whatever the initial draw, as we survey the band room of La Guardia High School, that loud, gleaming trumpet is all Dave's. Dave Guy loves playing the trumpet. But if we'd been talent scouts, standing in that basement in the mid-nineties, we probably would've walked right past him.

Dave's a junior, shy, a little pudgy, quiet when he's not playing. He's not the best trumpeter in the area. He isn't even first chair in his high school band—just one of countless other aspiring teenage musicians, patiently filling in that backup position. In 1995, he does not command attention. He is not a standout, not the next big thing.

Why are we telling you this story?

Today, many musical experts consider Dave Guy to be one of the greatest trumpet players alive. You may not know his name, but if you're a fan of

Jimmy Fallon's Tonight Show, you've definitely heard him. Dave Guy plays alongside The Roots, a critically acclaimed hip-hop group that also serves as Jimmy Fallon's house band.

When Eric Clapton needed a trumpet player for recording, he got Dave Guy. When Sharon Jones and the Dap Kings would go on tour, they called Dave Guy. He toured with Amy Winehouse. Jazz legend and nine-time Grammy winner Wynton Marsalis once gave Dave Guy a trumpet.

Guy is unquestionably one of the premier trumpet players out there today. So how did he get from that basement in La Guardia High School—where again, he really wasn't an early standout—to the Tonight Show?

Dave Guy succeeded by changing the wiring in his brain, something that every one of us has the opportunity to do. Up until fairly recently, we didn't know, on a technical level, what the map to excellence looked like. Today, we've aggregated enough information that we can give you a good overview of the route to achieving your long-term goal.

How does it work?

How did Dave Guy go from second chair to the main stage? How did a shy high school junior become one of the great trumpet players in the world today?

Dave Guy's story is in some ways not unique, but to understand it, we will first need to unravel some myths that lie at the heart of our very idea of genius.

The Roadmap to Success

When you look at people doing the best work in nearly any field, again and again you find that they have built very specific pieces of mental programming. And that sequence of brain apps is an accelerator for success.

They may not be aware of it. If their area of expertise doesn't involve brain science or psychology, they may have no idea what they've done to their brains. Yet research from all over the world is shedding light on a remarkable human evolutionary advantage called *neuroplasticity*. The brain has a remarkable capacity for change and improvement.

To be clear, this has nothing to do with that old, oft-cited chestnut about how we only use 10% of our brains. On the contrary, neuroscientists will tell you we do have access to the full landscape of our gray matter.

From a biological perspective, carrying around three pounds of brain is costly; maintaining your brain uses up to 20% of your body's oxygen, and 20% of its nutritional energy.[1] It also requires a pretty big skull to protect it, rendering human childbirth especially painful and—for most of our history—dangerous. In terms of evolution, it's hard to imagine lugging around a bunch of extra deadweight in our heads if only 10% is useable.

Neuroplasticity, the ability to wire and rewire your brain, is where the real action is. Your brain's physical ability to grow larger may be constrained by your skull; still, the flexibility of your brain's wiring system leaves plenty of room for forging a near-infinite number of connections, ideas, and thoughts.

Stephen Hawking, Neil De Grasse Tyson, and Sandra Day O'Connor didn't succeed thanks to the secret folds of their gray matter, but rather due to the trails they blazed by organizing and reorganizing their brain's neural wiring system.

Your neural wiring system is composed of neurons, chemical/electrical cells that connect with each other to transmit vital information from your senses, movement and thought. If neurons were Legos, the great thinkers of history would have some incredible structures on display— pieced together from these micro building blocks that they assembled in surprisingly new ways.

Your brain's neural wiring comes partially prewired at birth but it's also shaped by experience. From playing basketball to weaving baskets, you have the ability to program your brain for a whole host of behaviors or brain apps.

Think about the brain, unlike your smart phone, the "battery" runs for an average of 78 years, no electrical connection needed, nothing to plug in. You don't have to worry about it getting lost or stolen. Unless something has gone horribly wrong, you're never in danger of dropping it, or running it through the washing machine.

The technology will never become obsolete. No need to fret about whether it's time to buy the next upgrade. The brain is the ultimate flexible device.

And it doesn't matter if you want to become a great musician, a better gardener, the most successful computer programmer or salesperson—the solution is made up of the same small group of programs you need to create or reinforce. When put in the right sequence and properly habituated, these brain apps will create a unified map and accelerate your success.

The Seven Brain Apps

There are eight key factors in successfully reaching any long-term goal. Seven of them are within your control and can be turned into brain apps.

The eighth factor is luck.

Luck

We don't like to talk about the role of chance in our lives. It feels much better to believe that, through the dint of our own hard work, we will inevitably carve out our success. But of course, we can't shape everything about our lives.

There are three attributes to luck: you have to be in the right place, at the right time, under the right conditions. Sometimes we can tweak our situation to be more favorable.

In *Thinking, Fast and Slow*,[2] Nobel Prize winning scientist Daniel Kahneman sums up the importance of luck with the following two equations:

> **Success** = hard work + luck
>
> **Greater success** = a little more hard work + a lot of luck

Even if your brain is perfectly primed, you still need to be in the right place, at the right time, under the right conditions. If Michelangelo hadn't lived during the Renaissance, would the world have the statue of David? Astrophysicist Cecilia Payne-Gaposchkin discovered what the universe is made of; could she have cracked the code if she'd been born to 17th century Puritans?[3] If Paul McCartney hadn't met John Lennon in 1957, what would have happened to the history of rock and roll? (Ringo's life would have been pretty different, for starters.)

The Talent Myth

For many of us the story goes something like this: talent is a gift, bestowed upon you through luck—generosity from the gods, or perhaps your genes. It's an unspoken assumption; normally we don't even think of questioning it. 'Here are the things I'm naturally good at, here are the things I can't do.' And when we think about the most brilliant achievers out there, the geniuses of whatever field you want to mention—Bruce Springsteen, Carl Sagan, Meryl Streep—we assume they have some special dose of pixie dust that us mere mortals can't access.

That talent lottery is fundamental to how many of us see the world. The only problem with this thinking is that it's flat-out wrong. It's not the first time a culture has been mistaken on this level: until the late 19th century, most Westerners believed sickness was caused by breathing bad air. As late as the 20th century it was widely thought tuberculosis was the result of "poor moral character."[4]

In the same way germ science turned medicine upside down, new research is rewriting everything we thought we knew about talent. When we take a closer look at everyone from Michael Jordan to Albert Einstein to J.K. Rowling, there is nothing about their early lives to hint at some innate seeds of greatness. In fact, sometimes it's just the opposite.

Michael Jordan failed to make his varsity high school basketball team and instead was placed on the junior varsity squad. The story goes that when he got home, young Michael locked himself in his bedroom and cried.

Albert Einstein graduated from college with an overall lackluster academic performance. He tried repeatedly to get a teaching position and eventually was forced to take an uninspiring clerk job in a patent office.[5]

JK Rowling was a divorced single mother, living on public assistance. Twelve publishing houses turned down her Harry Potter stories. She was advised to keep her day job; the only problem was that she didn't have one.[6]

But how can skills not be inborn? Can't we see inherent differences in abilities, even among kindergartners? After all, it is true that if you went down to a hockey rink and watched a group of five-year-olds skate, you could absolutely identify that some kids were better than others. You would say, "Well, that kid scoring the goals, he's got talent."

But what we've discovered is that this early advantage is not a predictor of future accomplishment. And even when these young achievers do well later, it may have less to do with some innate traits and more related to the extra attention and support they receive once adults have labeled them as 'special.'

For instance, in *Outliers*, Malcolm Gladwell points out that a disproportionate number of Canadian hockey stars were born in the first half of the year. Gladwell suggests a simple explanation, arguing that it's not, say, a matter of Zodiac signs.[7] When they started playing hockey as small children, they were a little stronger and faster than their slightly younger peers, and so they received more training. This training, in turn, gave them an extra edge, and the effects compounded.[8]

Grand Valley University professor Robert Deaner read this and did a study of his own.[9] Surveying 2,700 players of the NHL, he found that 35% were born in the first three months of the year. A disproportionate amount of hockey players really had started out as the older kids in their junior hockey teams. But Deaner found something else interesting: as a group, they only scored 25% of the goals.

Meanwhile, the players born in the last three months scored 44% of the goals. That initial advantage might have gotten athletes more attention from coaches and scouts, but it seems that frequently the greatest players were motivated by what Deaner calls the "underdog effect"; a childhood spent competing against older, bigger kids left them with an incredible determination and work ethic. Two-time Olympic gold medalist Mia Hamm honed her early skills chasing after her older brothers on the soccer field, constantly pushing herself to surpass kids taller and stronger than her.

There are a lot of people who display early talent and never amount to anything in their field. There are plenty of people who show no talent in the beginning and go on to become incredibly successful. Initial talent is not the main determiner of future success. So what is?

Setting Your Mind Up for Success

For our first brain app, we turn to the work of Stanford's Professor Carol Dweck. Dweck has spent decades researching achievement and success. She describes her groundbreaking discoveries in her book *Mindset, the New Psychology of Success*.[10]

Dweck explains that our quality of life is greatly influenced by our framework for seeing the world: mindset. There are two basic types of mindsets available to us: growth or fixed.

Growth Mindset

Growth mindset frames skill as something you develop throughout your life. Everything from your intelligence to your ability to win at foosball is malleable and largely the result of how much effort you're willing to put in.

Because of the emphasis on learning more and pushing yourself, a growth mindset embraces challenges and persists in the face of obstacles. After all, why be afraid of making mistakes, when messing up at first is a natural part of the process? These setbacks aren't seen as a sign of weakness or failure, but as learning experiences.

In growth mindset, you welcome constructive criticism as an opportunity to learn what to focus on for improvement. The success of others is inspirational; if they can do it, so can you. They help establish the path to mastery for you to follow.

Fixed Mindset

Fixed mindset frames skill as fixed or limited. Everyone is born with a certain finite amount; there's nothing you can do to affect it. You can only hope you're one of the lucky few, and try to limit what challenges you face, since making mistakes is surely a sign you were never blessed to begin with.

Obstacles are terrible risks for further mistakes. When you encounter them, you're less likely to put forth your best effort. You may even hedge against failure by implying you weren't really trying in the first place.

Criticism is very threatening. Your options are to resign yourself to your unfixable shortcomings, or turn on the person giving the critique, dismissing their suggestions.

Other people's successes make you uncomfortable, since they only serve to highlight assets you may not have. Watching someone else slip up provides some small measure of comfort. You may even find yourself rooting for them to fail.

You tend to be overly hard on yourself, often holding yourself to unachievable goals and standards. You're cautious of new experiences because the fear of failure increases with unfamiliarity. Your comfort lies in doing things the way you've always done them. Change is the enemy.

Fixed Mindset in Business: A Case Study

Can an entire company have a fixed mindset? Absolutely, says Carol Dweck. If those in charge push their attitudes hard enough, it can trickle down through an entire institution—with disastrous results.

Hey, remember BlackBerry?

It's a distant memory now, but less than a decade ago, they were at the technological forefront of handheld devices. The BlackBerry was revolutionary—first as a two-way pager and later as a cellphone.

In a time when cellphones weren't good for much beyond making calls, Canadian tech company RIM (Research In Motion) had created a palm-sized PDA that could send and receive e-mails from anywhere. The network was secure, the battery lasted seemingly forever, and the little QWERTY keyboard let you tap out a message with nearly the efficiency of typing on a computer keyboard.

In the immediate aftermath of the September 11th attacks, phone networks were swamped with calls, and the BlackBerry network was one of few that stayed operational. Vice President Dick Cheney worked his BlackBerry feverishly in those crucial hours. The US government bought 3000 BlackBerry phones and distributed them to elected officials and staffers all over Washington, according to Alastair Sweeny in his book *BlackBerry Planet*.[11]

From Wall Street to Capitol Hill, BlackBerry smartphones became the must-have status symbol of the busy, the important, and the with-it. Al Gore wore his on his hip. Users joked they were addicted, to the point where Webster added "CrackBerry" to the dictionary as their 2006 Word of the Year. By 2009,[12] RIM was named by *Fortune* magazine as the fastest growing company in the world, with earnings spiraling at an unprecedented 84% over three years.[13]

The two men behind the success of the BlackBerry were founder and inventor Mike Lazaridis and his co-CEO, Jim Balsillie. Unfortunately, they would also be behind its fall from grace.[14]

The first hints of trouble had already surfaced in the summer of 2007, when RIM still appeared to be on top of the game. Apple introduced its first iPhone, with a new touch screen. Many in the tech industry stopped and took notice. The RIM CEOs were not among them. As late as May of 2008, Lazaridis was still brushing the touchscreen aside, insisting, "The most exciting mobile trend is full QWERTY keyboards. I'm sorry, it really is. I'm not making this up."[15]

(He was making this up.)

It wasn't the first time Lazaridis had failed to appreciate the true technological potential of the devices he'd helped to pioneer. When PDAs in Asia began to offer color screens, he suggested it was just a fad. "Do I need to read my e-mail in color?" he's reported to have said.[16]

Maybe in another company, someone might have stepped forward and delivered a wakeup call. But Lazaridis was notorious for only hiring people who agreed with him; he avoided dissent.[17] Trapped in an echo chamber of their own design, Lazaridis and Balsillie continued to insist their practical, workmanlike product had an impossible-to-beat foothold among businesspeople. How could a phone that wasted battery life on shiny new features elbow in on their territory? Who would tolerate the less user-friendly touchscreen keyboard on an iPhone?

Well, since you're reading this in the present, you already know the answer.

That oh-so exciting built-in keyboard became a shackle, cutting possible screen space in half and severely limiting other options the BlackBerry could offer. Meanwhile, more and more consumers were enticed into the world of smartphones by the bells and whistles of the new apps. The iPhone had altered the very definition of what a phone was supposed to do.

By the time even Lazaridis and Balsillie could no longer deny that change was needed, it was too late: they'd lost their edge, their voice of authority. When they finally started to offer their own touchscreens, it came with a feeling of sweaty desperation—and amazingly, their first attempt at an iPad competitor didn't even offer email.

Over a four-year period, RIM went from an unbelievable 50% of the smartphone market to just 1.5%.[18] Very few companies survive this kind of reversal of fortune and live to tell about it.

Forbes magazine lists Thorsten Heins as one of the worst CEOs of all time. Heins headed up RIM from January 2012 to November of 2013.[19] While it's true he was unable to send the BlackBerry soaring back to its old heights, it's also true that Lazaridis and Balsillie handed him the reins to a company they'd already driven straight into the ground.

How had Lazaridis and Balsillie gotten things so wrong? It's easy, with the benefit of hindsight, to laugh at their mistakes. Who looks at the smartphone market of 2008 and names built-in keyboards, which RIM had been offering for nine years, as the most exciting innovation? Who fails to anticipate that PDAs, following the trend of every visual medium from photographs to computer screens, might someday switch to color?

The fact is, both CEOs were bright, talented people. Certainly, they had the know-how to get on top of the industry in the first place. They set out to build a handheld e-mail sending device, and they built a great one. RIM's problem wasn't idiocy; it was fixed mindset.

Fixed mindset results in fearful, risk-minimizing behavior, but it's not the same thing as humility. It's possible to have a fixed outlook while still behaving in dangerously arrogant ways.

In their minds, BlackBerry was fundamentally a winning product, and they were fundamentally winners. They had scaled the highest peak they could find, and once there, they had no interest in seeking out other mountains, in branching out and exploring what a smartphone could be. They got complacent.

The iPhone wasn't a wakeup call to Lazaridis and Balsillie; it was a threat. Because they couldn't risk admitting to themselves that maybe change would be necessary, they chose to bury their heads in the sand.

RIM continued to produce full keyboard phones long after consumers and even their own internal research had made it clear that touchscreen was becoming the market's preference.

Dweck says in fixed mindset mode you spend your time "trying to prove, rather than improve."[20] For example, Lazaridis and Balsillie weren't interested in continuing to innovate. Their energies were spent putting out sound bites about the superiority of QWERTY keyboards.

Apple and fellow competitors capitalized on RIM's fixed mindset. They aggressively pushed the smartphone platform to new heights, adding a plethora of apps and features, making them more computer than phone. Email had been just the beginning. RIM found itself playing catch-up, and in the fast-paced and unforgiving world of technology, that is never where you want to be.

The Anatomy of Fear

If fixed mindset is so detrimental, why do we have it? Where does it come from?

To understand the answer, we need to journey back about 40,000 years to the point in time when our ancestors were roaming the great savannas, alternating between chasing prey and being prey. Their systems had already adapted to face the harsh environment, according to Rick Hanson PhD, coauthor of *Buddha's Brain: The Practical Neuroscience of Happiness, Love, and Wisdom*.[21]

Today, those ancient adaptations show up in the most peculiar ways as we constantly hijack a system built for times long past. The prospect of needing to radically redesign your smartphone might not seem an awful lot like being chased by a lion, but to the brain's subcortical structures, it's pretty much the same deal.

Hanson explains how it all works. As you let yourself consider everything that could go wrong with abandoning your QWERTY keyboard, your brain sounds the ancient 'lion alarm,' and begins to prepare for battle. First, stress hormones like epinephrine kick your heart rate up and norepinephrine increases blood flow to bring your largest muscles online faster. Your pupils dilate to take in more light for enhanced visibility, and your bronchioles expand, boosting lung capacity for fight or flight.

Another stress hormone called cortisol jumps in to suppress your pain receptors, just in case you happened to be wounded. The hippocampal system normally does its part to keep the cortisol level of the amygdala (the brain's fear center) under wraps. Under stress, it takes a step back and lets the amygdala ratchet up, driving more cortisol into your system, in effect supercharging your blood, much like adding octane to your vehicle's gas tank.

The system that governs reproduction essentially turns off, because that's not a priority at the moment; and your digestive system also goes into hibernation, allowing the body to redirect energy and blood flow elsewhere as needed.

All of this signals your amygdala to go on a higher level alert; this system, which normally monitors threatening information, turns up the heat on your emotional thermometer and moves the needle from stress to fear or anger, fueling you for a life-and-death struggle.

Your prefrontal cortex—the home of reason, speculation, planning and assessment—gets hijacked by more primitive systems like the amygdala, essentially turning the keys to your body over to your reptilian brain. There's not always time to appreciate that your competitor may be onto something when you're in "kill or be killed" mode.

There was a time when the primitive elegance of this system made sense. But that was thousands of years ago. Today a whole host of activities, from getting cut off on the road to receiving a bad email from your boss, to a significant challenge by a competitor's product, can trip the same ancient survival system.

Sadly, for all too many people, there is a metaphorical lion lurking around every corner. The result is a misplaced sense of fear. That's how you can have a razor-sharp intellect and still fall prey to fixed mindset. It has nothing to do with intelligence; it's about fear stepping over the line of rationality to dominate your decision-making.

Many people find it difficult to deactivate the stress switch. Hanson suggests practicing mindfulness and meditation to keep the brain from prematurely sounding the alarm bell all day long. (More on this later in Chapter Seven.)

We've seen an example of fixed mindset in action.

Now, what does growth mindset look like?

The Growth Mindset Town

Sitting on the north bank of the Tennessee River is an unlikely success story, almost a perfect illustration of just how much can go right when growth-minded attitudes are in place.

Muscle Shoals in the 1960's was a sleepy, bucolic Alabama town. There was no nightlife. For a young person, there were few distractions, few options for channeling restless energy. When we spoke with her, *Muscle Shoals Sound Studio* author Carla Jean Whitley, summed it up best: "You could get into trouble, or you could get into music."[22]

In 1965, songwriter and musician Rick Hall opened up a recording studio in Muscle Shoals he optimistically called FAME Studios. Hall grew up in abject poverty and was driven to make a name for himself. Armed with more determination than money, he recruited young unknown local players as his studio musicians, most likely because they were available and—more importantly—affordable.

Ultimately, Hall's lineup featured Roger Hawkins on the drums, guitarist Jimmy Johnson, Barry Beckett on piano, and bassist David Hood. All four men were born and raised in close proximity to Muscle Shoals. Their musical style reflected what was available to them—the songs they heard on the radio, the records they could buy, and the local music of the area, a rich mix of old time folk, country, and rhythm & blues.

Desperate for success, Hall set his early expectations high. Some, like producer Jerry Wexler, would later characterize him as overbearing and tyrannical, but he did get results. The Muscle Shoals Rhythm Section, or "Swampers" as they came to be known, worked hard and put in long hours.

Hall's first big break came in 1966, with Percy Sledge's soon-to-be classic "When a Man Loves a Woman." At the time, Sledge was working as an orderly, singing to patients on weekdays and touring with the Esquires Combo on weekends. "When a Man Loves a Woman" put him—and Muscle Shoals—on the map.

A bunch of local boys had scored their first big hit, but instead of getting overwhelmed or complacent, the Swampers vowed to bring their very best to the next session. They didn't overthink it; they showed up ready to create another honest musical experience.

The Swampers were young and inexperienced. They recognized just how much they didn't know. At the studios of Memphis, Nashville, Chicago, and New York, the industry standard was for musicians to work off of pre-written musical arrangements. But not all of the Swampers could even read music; they looked at chord charts and went by feel.

Dweck points out that people with a growth mindset are much more likely to assess their talent accurately. It's far easier to make an honest judgment when your ego and very sense of self-worth aren't tied up in the verdict. This understanding is crucial; you need to recognize where you're at before you can plan the steps that will take you where you want to go. And the Swampers were ready to take those steps.

Working off chord charts and instinct left plenty of room for improvising and experimenting, and that flexibility was part of the key to their success. The Swampers were constantly growing as musicians. Whenever a recording artist arrived in the studio, it was a new challenge: how could they best serve the new unique song? What playing style would make the singer shine? They became the ultimate chameleons, mastering an incredible array of genres.

Each piece of music and each artist seemed to unlock another idea. They went through these opportunities, discarding what didn't work, and hanging on to what did. They were in effect being paid to learn, and the hit recordings affirmed that they were on the right track. It was a critical feedback loop.

Eventually, the Swampers would leave FAME and open up their own recording venture, Muscle Shoals Sound Studio. They brought the same dogged determination and tailor-made approach to the work at their new location, a converted blinds factory at 3614 Jackson Highway.

Their first album, Cher's aptly-titled *3614 Jackson Highway*, was a critical success but a commercial disappointment, peaking at only 160 on the charts. But the Swampers didn't despair. Worrying that you could lose "the gift" is, of course, a hallmark of a fixed mindset.

"The sense that I got from Jimmy Johnson was, it kind of put them on their toes," says author Whitley. After years of focusing solely on the music part of the industry, it was time for them to think about the industry.

Whitley summarized their new attitude. *"'Okay, there are different things we need to think about, from a business perspective. Everything's not going to be a hit, and it's not going to be a hit the second we finish recording it.'* There's a lot more involved than just showing up and creating the work...and definitely they learned a lot about that, very quickly."

The Swampers left their mark on music history. Look back through the iconic hits of the 1960's and 1970's and the Swampers are everywhere, playing with Aretha Franklin ("Respect", "Chain of Fools", "I Never Loved a Man"), The Rolling Stones ("Brown Sugar"), Wilson Pickett ("Mustang Sally"), Etta James ("Tell Mama"), Paul Simon ("Kodachrome", "Loves Me Like a Rock"), The Staple Singers ("I'll Take You There"), Bob Seger ("Old Time Rock and Roll"), and Lynyrd Skynyrd ("Free Bird"), among many others.

The risk of striking out on their own definitely paid off, sometimes in new and exciting ways. "They were able to get a lot more writing credits, and production credits," Whitley notes. "The Swampers really wanted to move into producing. Well, by virtue of owning the studio and being the guys who had to make it work, they stumbled into more producing opportunities." When the Rolling Stones came to town, Jimmy Johnson found "they just sort of assumed he would step into that role. So he did."

They made such an impression on Lynyrd Skynyrd, they famously earned a shout-out on 1974's "Sweet Home Alabama."

With their hallmark 'work it out as you go' style, the Swampers enshrined themselves and their sound in the canon of popular music.

As noted earlier, when he was a junior in high school, Michael Jordan tried out for the varsity team and didn't make the cut. Stories like this are sometimes told as an invitation to look down at the foolish coaches or other authority figures who failed to spot the young superstar in their midst. The truth is, if Michael Jordan had walked into that gym with Bulls-level talent; he would've made the team in a heartbeat. He wasn't quite there yet.

What set him apart from other disappointed teenage hopefuls was that, once he'd had a good cry, Jordan took the rejection as a personal challenge and worked hard to raise the level of his game. He became legendary for the intensity of his practice regimen, and the results speak for themselves.

Growing Pains

Remember Dave Guy, the average high school trumpet player? Now, Dave had a stroke of good luck: he lived in New York City. Live music was everywhere.

And so, after high school, Dave went around playing with any band that would allow him onstage. That meant playing any style: one night it would be rock, then blues, then salsa, then soul. Like the Swampers from Muscle Shoals, Dave gained a wide background and knowledge in his craft.

The palette of ideas he was working from, the number of different techniques available to him, kept expanding, and by recombining those ingredients, he could start to come up with some new, exciting ideas. Frans Johansson calls this the Medici Effect. It's one of the keys to creativity and we'll go into this more in Chapter Six.

But in the meantime, there was young Dave Guy, playing unfamiliar styles. When we interviewed Dave, we asked him, "So you're a hip hop guy, and you're playing some other genre of music. How did it feel to be out there performing, knowing you weren't the best player in the room?"

And it wasn't just that; he made it clear that the other musicians had no problem razzing him when he slipped up. To someone with a fixed mindset, that kind of negative feedback could have been the end.

We wondered if, after a session of verbal hits from the veterans, he ever left vowing to stick to hip-hop.

The thought had never even occurred to him. "It was part of me paying my dues," he told us. "My whole goal was just trying to get better, and the only way I was going to get better would be by practicing with better players." Dave was completely willing to risk screwing up; in his mind it was a necessary step for getting where he wanted to be.

The ability and willingness to fail in the learning phase is one of the keys to achieving and raising yourself up to the next level.

Growth mindset embraces challenges, sees obstacles as something to overcome, and sees mistakes as an opportunity to improve. Think back on the major learning experiences of your life, the ones that have really stuck with you. They probably involve making mistakes.

Learning from Failure

When you think of failure, it's unlikely the prodigious inventor Thomas Elva Edison comes to mind. After all, he gave us the electric light bulb, which literally and figuratively changed the way we see the world.

By the time the wizard of Menlo Park died in 1936 at the age of 86, he had managed to attach his name to 2,332 patents worldwide.[23] Some of those creations he produced personally, some he helped develop as part of a team, and sometimes the connection is dubious. It seems that if you worked in Edison's New Jersey laboratory, he had no problem sticking his name on your work. (It apparently helped drive Nikola Tesla from the building, along with Edison's unreliable approach to paying him.)[24]

Still, there is no denying that Edison had deep domain knowledge. The light bulb, the phonograph, and the movie camera— these are the kinds of triumphs we associate with someone operating at the top of their inventive game.

Edison embodies the very essence of growth mindset in his famous quote, "Genius is one percent inspiration and 99 percent perspiration."

The mythmaking process tends to edit out a person's missteps. We may remember his legendary persistence in trying filament after filament, as he searched for a material that could sustain the incandescent glow of electricity, but the punch line is still unmitigated success.

Edison was not, however, immune to all-out failure. Writing for Smithsonian.com Erica R. Hendry recounts some of his more notable missteps.[25]

For instance, what do you do when you hold the corner on sound recording technology? Edison decided to team up with a German doll manufacturer to produce a talking doll. If you're picturing a reassuring Tom Hanks as Woody from Toy Story, think again. According to Leonard DeGraaf, an archivist at the Thomas Edison National Historical Park, the voice leaking out from Edison's girl doll was 'just ghastly.' It wasn't just a creepy doll; it was a malfunctioning doll, fragile and prone to breaking. After 30 days, Edison recalled her from store shelves.[26]

Edison also invented a vibrating 'electric pen' designed to puncture paper, creating a dimpled template, which would then serve as the master pattern for reproducing additional copies. The final design was awkward, heavy, noisy, messy and difficult to control.

Attempting to recoup some of his loss, Edison sold the technology and patent rights. A. B. Dick bought a piece of that technology, reconfigured it, and created the first mimeograph machine, thus becoming for many years the hero of churches and grade school teachers everywhere. Although it's harder to trace, some say the tattoo machine also owes its roots to Edison's abandoned contraption.

Putting aside Edison's less than honorable treatment of employees, collaborators and competitors, maybe the most impressive part of his story is the sheer number of times Edison tried and failed.

He lost hours, days and years in pursuit of his goals, often ending broke and physically beaten down. Deaf and diminished, still he persevered to his dying day, choosing to focus on what might lie ahead, instead of past victories and the many missteps that punctuated his long inventive career.

You might think about that tonight when you flip off the light switch, provided you can get his creepy doll out of your head.

Like Dave Guy and so many others, Edison demonstrated that growth mindset allows you to overcome or at least negotiate obstacles, and keep moving forward.

We've seen that if you have a fixed mindset, you regard traits like inventiveness, skill and intelligence as inborn, and try to live your life either convincing yourself and others that you have them, or hiding the fact that you don't. This leads you to avoid challenges. Complacency feels more comfortable and change becomes something to avoid. You beat yourself up over your mistakes and dwell on them.

If you have a growth mindset, you understand that qualities and skills can be cultivated with effort. Anyone can grow and change through experience. Growth is what generates interest and excitement, but often involves rough patches along the way, as even a great inventor like Edison could attest.

Most of us exhibit a complex blend of fixed and growth mindset in our lives. However, culturally, we seem to tend toward a more fixed perspective. Would you rather be told, "Wow, you're a natural", or "You must have worked really hard at that"? To many of us, being praised for effort alone feels like a backhanded compliment. We love the story of the young hotshot who comes in and blows away the by-the-book hard worker with a blaze of brilliant natural talent. We see it everywhere from sports to action movies, from academia to the arts.

CHAPTER 1

In her groundbreaking essay "Why Have There Been No Great Women Artists?"art historian Linda Nochlin outlines a familiar and common legend: an established painter stumbling upon a young "Boy Wonder," whose seemingly star-touched talent has arisen from thin air. "Usually," she notes, "in the guise of a lowly shepherd boy". She lists seven prominent artists, including the Renaissance's Giotto and Spanish Romance painter and printmaker Goya, all of whom "through some mysterious coincidence," happen to share the same highly specific origin story.

"Even when the young great artist was not fortunate enough to come equipped with a flock of sheep," she notes wryly, "his talent always seems to have manifested itself very early, and independent of any external encouragement..."[27]

With wit and common sense, Nochlin explains that many, many hours of instruction and hard work are required before anyone develops the technical skills to paint a convincing human figure. She argues that part of what set the great artists apart was not some divine spark but instead access to years of training and dedicated practice.

This, she says, explains why the golden age of classical painting yielded so few women artists—women just didn't have access to the opportunities they would've needed to hone their skills. Male apprenticeship programs, on the other hand, were natural breeding grounds for talent: young men gathered together, devoting all their time to the study of art under a highly regarded master artist.

In retrospect, it seems obvious: the execution of visual arts requires enhanced muscle control, and nobody is born wired to render all the details of the human form. Like a golf swing, painting must be perfected over time.

Still, whenever we recognize a tremendous talent, we are quick to dust off those tales of the Boy Wonder, brushing aside an individual's background and years of hard work. Fourteen-year-old Pablo Picasso passed the entrance exam to the Barcelona School of Fine Arts in just one day, but while that is extremely impressive, Nochlin reminds us that Picasso's father was an artist and an art professor. By fourteen, young Pablo had already received years of formal training.

This is a far less appealing story. As Dweck points out, even in the parable of the tortoise and the hare, nobody really walks away vowing to be more tortoise-like. It's the false dichotomy of being either a fascinating genius, or dull but virtuous.

The need to apply real effort can be scary. It casts doubt on your own innate brilliance, and it removes possible excuses if you do fail. Culturally, It's more exciting to imagine people endowed with special magical powers.

But if Dave Guy had a magical power, it was his willingness to learn and sometimes fail as he worked with better players. The Swampers, playing in so many different genres, were constantly forced into new learning experiences as well.

As Dweck explains, fixed mindset is outcome-focused, but ironically, growth mindset produces more favorable outcomes. Growth mindset engenders teamwork, not competition, because the abilities aren't threats but learning opportunities. The approach becomes, 'if they can do it, so can I'. This concept is a major driver in accomplishment. It's sometimes known as the Bannister Effect.

The Bannister Effect

Over sixty years ago, conventional wisdom held that running a sub four-minute mile was impossible. After all, human endurance had its limits. The lungs and heart could only produce so much blood-rich oxygen, and the muscles could only metabolize what the lungs and heart could deliver.

For many years, athletes had tried to defy the presumed laws of human mechanics and squeak out a mile in less than 240 seconds. The net result was always the same: defeat.

Then on May 6th, 1954, a tall unassuming lad from Great Britain named Roger Bannister came along and blew the impossible away with a time of 3 minutes, 59.4 seconds.[28] It made international news. He became an overnight sensation.

But here's what's interesting: although runners had tried and failed to clock that time for years, within just 46 days, Bannister's feat had been duplicated. With fair regularity, the record time would continue to drop.

So what happened? Why was there a rash of sub four-minute milers after Bannister, when for years no one seemed capable of either starting or joining this elite 239-second club?

Athletes in 1954 did not undergo some fundamental physical transformation. It's now clear that many runners actually had the raw ingredients necessary to break the record. What these four-minute wannabes lacked was the ability to believe the goal was truly obtainable. The barrier was mental. In short, it was an issue of mindset.

This gulf between our capabilities and our mental assessment is one of the most profound disconnects in the human experience. When people achieve beyond what we thought possible, we tend to elevate them above the rest of us, wondering what special quality allowed them to break those boundaries. We don't always examine what caused us to believe in those boundaries in the first place.

Bannister had a growth mindset. He saw the goal as obtainable. His combination of preparation and perspiration is a basic winning formula for skill acquisition across a wide variety of enterprises, from math to music to athletics and beyond.

But in some ways, the real heroes are the ones who followed in Bannister's running steps. These athletes continue to push at the barriers for themselves—and for the rest of us.

"*Nanos gigantum humeris insidentes*," as the medieval saying goes. We're standing on the shoulders of giants. It's much easier to see the possibilities of what lies ahead by capitalizing on the previous work of others.

When enough people do that, a barrier breaks under its own weight. The current mile record is held by Moroccan two-time Olympic gold medalist and "King of the Mile" Hicham El Guerrouj, at 3 minutes 34.13 seconds.[29] That's a whopping 25 seconds faster than Bannister. El Guerrouj has held the honor since 1999, but if history is any predictor, it's only a matter of time until the king is dethroned and the record is broken again—and again.

How can we make sure we maintain forward momentum?

You can coach people toward a growth mindset by focusing praise on their effort and hard work. Conversely, you can unintentionally push people in the direction of a fixed mindset by praising their supposed inborn abilities. This often has the unintended consequences of creating a ceiling for their improvement.

People with growth mindset like Roger Bannister, Dave Guy and the Swampers view setbacks as educational, a call to action. They take charge of their own learning processes. They don't blame their failures on others.

Mindset in Coaching and Beyond

Bill Bowerman's name may not be a household word, especially since he died more than a decade ago. But if you've laced up a pair of gym shoes in the last 44 years, he's probably had at least some impact on your life. Or, more literally, he's reduced some of the impact on your feet.

In Portland, Bowerman is recognized as an extraordinary University of Oregon track coach, whose career spanned from 1948 to 1973— and produced 16 sub four-minute milers, 24 NCAA champions, 12 American record holders, 51 All-Americans, and 31 Olympic athletes.[30] Bowerman's most famous running student was the legendary middle-to-long distance runner Steve Prefontaine.

In the early 70's, Prefontaine was a highly recruited high schooler who chose the University of Oregon specifically to train under Bowerman and his cutting-edge program.[31] The young phenom was surprised when Bowerman initially replied to his query with the standard boilerplate U of O recruitment form letter.

Bowerman explained later that he was very eager to have Steve on the team—he'd followed his high school career with a great deal of interest—but thought it inappropriate to show favoritism. Although Bowerman was undoubtedly being truthful, what is not stated, was the underlying power play in sending one of the great runners in the country a form letter.

While Prefontaine was being lavishly courted by every major running school in the country, Bowerman sent a clear growth mindset message that his program wasn't about celebrity status; regardless of prior success and accolades, everyone was expected to walk on the track ready to work, learn and contribute. Hard work and incremental improvement drove the gears of success. Bowerman was delivering the classic growth mindset lesson to his young protégé.

The lesson took hold. Under Bowerman's tutelage, Prefontaine never lost a home turf race in distances greater than a mile. He was, however, no slouch in the mile either, with a career best of 3 minutes, 54 seconds—a time that, in the Roger Bannister days, would have decimated the world record. Prefontaine once held seven American records, in distances ranging from 2000 to 10,000 meters.[32] Tragically, his career was cut short at the age of twenty-four by a fatal car crash.

As a coach, Bowerman was renowned for his growth mindset. He was always on the lookout for any legal advantage he could bring his runners. From early experimentation with hydration drinks, track surfaces, equipment, and running shoes, he was tireless in his drive to push the outer edges of running technology.

In 1962, he came back from a trip to New Zealand with a new fitness regimen. It was called jogging. It's hard to believe, but the concept of jogging for recreation and health was a foreign concept in the United States as late as the Camelot days of the Kennedy administration.

Bowerman made it his mission to promote the benefits of jogging in the U.S. and, in 1967, along with cardiologist W.E. Harris, he wrote a pamphlet called *Jogging*.[33] It sold over a million copies. It is credited for fueling the jogging revolution that helped transform the American concept of exercising.

If a key aspect of growth mindset is the willingness to risk, then Bowerman can check that box.

In January 1964, he and one of his former running stars at the University of Oregon, Phil Knight, noticed a dearth in performance running shoes. Up until that point, what shoes there were had largely been coming out of Germany.[34]

Bowerman and Knight each put up $500 and, on a handshake, created a running shoe startup named Blue Ribbon Sports. It was built on the back of Bowerman's ideas for a whole new class of high-performance American running shoes.[35]

In the early days, Bowerman, like so many entrepreneurs, drew his inspiration from unlikely sources. The story goes that he had an epiphany one morning at the family breakfast table, over a plate of waffles.

He realized that if the grid pattern from the waffle iron was incorporated in the bottom of a running shoe, it would offer the runner both a lighter shoe, due to a reduction in surface area, and one with better gripping power as a result of the ridges and valleys. This had been a standard traction feature in automobile tire design for years.

By 1964, the waffled sole prototype had been built and tested, and Knight began pedaling these new-fangled lightweight solid grip running shoes out of the trunk of his car. As early as 1965, business was strong enough that the duo felt the need to add a third partner. Running enthusiast Jeff Johnson joined Blue Ribbon Sports that year.

Johnson handled everything from collateral to distribution to retail. In his spare time, he even worked on new shoe designs of his own. With Johnson's dedicated work ethic and marketing prowess, the company took off.

In 1971, with a desire to keep the brand moving forward, the trio sought the aid of a design student from Portland State University named Carolyn Davidson for a new logo identifier. She obliged and the now ubiquitous "swoosh" was born, along with a new company name: Nike.

Since 1971, Nike has been at the forefront of running shoes and apparel, garnering international respect and recognition. Bowerman died in 1999 at the age of eighty-eight. His legacy lives on the feet of millions of runners all over the world, from the professional ranks to weekend warriors.

Bowerman is quoted as having said, "If you have a body, you are an athlete."[36] We humans, his words seem to imply, have an almost limitless potential. Achievement depends on your ability and willingness to frame your experiences, proactively, in growth mindset.

As a coach, Bowerman offered constructive criticism based on a balanced observational approach. He was always interested in figuring out what he could do to help runners achieve their best. This ran the gamut from personal growth to performance equipment. If a runner did his best and the team still came up short of victory, then so be it. Effort did not always guarantee a win, but it was the way Bowerman measured success. This strategy contributed to his standout coaching record. Bowerman carried this same approach into business, where his passion for exploring new possibilities seeded a level of innovation that still reverberates through the Nike culture today.

Mindset and Performance in Business

But not everybody thinks like Bill Bowerman. When it comes to mindset in business, the so-called "CEO disease" is a fatal flaw characterized by avoiding failures, risks, and the hard, long-term practice and training which ultimately benefits a company over time.[37]

It's fair to say that BlackBerry's co-CEOs Mike Lazaridis and Jim Balsillie developed a bad case of CEO disease. Ironically, when they were building their organization, their tolerance for risk was much higher. Once they achieved success, their mindsets became more fixed. The net result was devastating. BlackBerry's stock went from $230 a share in the halcyon days to as low $10 when their competitors overtook them.[38] In September 2016 new CEO John Chen announced that Blackberry was getting out of the phone manufacturing business, and instead would concentrate on software.

Growth mindset CEOs can turn failing companies around:

IBM once had a culture of arrogance and frequent turf wars. When CEO Lou Gerstner took over, he talked to every employee he could about what could be done better.[39] He attacked elitism, based executive bonuses on overall company performance, rewarded teamwork, put the customer first again, and rescued IBM.

Dweck says Anne Mulcahy did similar things at Xerox.[40] When she became CEO, the company was massively in debt, failing, and saddled with the same toxic culture as pre-rescue IBM. Mulcahy dug in to understand every aspect of the business, no matter how inconsequential it might have seemed—truly a growth-minded approach. She dealt in hard truths, but also cared deeply about morale and her employees' personal development.

Fixed mindset CEOs like Mike Lazaridis and Jim Balsillie create an environment vulnerable to groupthink. If you tend to hire people that think like you, your organizational strategy can stagnate. Although they added some features as time went on, they failed to understand the wider implication of what a handheld computer might offer to consumers. Their competition didn't make that mistake.

To guard against strategic stagnation, Dweck offers up six key attributes found in healthy companies, and all are consistent with the growth mindset framework:

1. Hiring partly based on enthusiasm for learning

2. Welcoming honest feedback

3. Presenting skills as learnable

4. Valuing learning and perseverance over inborn "brilliance"

5. Giving feedback to promote future improvement

6. Presenting managers as resources for learning

Mindset Pitfalls

Always be careful not to apply labels to yourself: "This happened, so I'm a loser." This is what Rick Hanson in *Buddha's Brain* would call secondary darts. Hanson asserts that much of human suffering is avoidable. It's not the first dart, the actual setbacks that get us down, it's the way we torment ourselves about them, the self-created secondary darts. We fixate on what doesn't live up to our expectations, slinging blame and turning failure from a learning opportunity into an identity, a condemnation.

Some people cling to a fixed mindset because at one point in their lives it made them feel affirmed, smart and talented. It gave them what, at least temporarily, felt like a route to self-esteem. It feels good to be labeled "talented." But these labels just as frequently limit us, barring us from true self-improvement.

Scott Ainslie literally wrote the book on famed bluesman Robert Johnson, *Robert Johnson at the Crossroads: The Authoritative Guitar Transcriptions*,[41] and is also a well-respected musician in his own right. He once told me that he used to get annoyed when people would praise him for his "natural gift."

It's not that Scott didn't appreciate the kind words, but he was very much aware that his playing was not something he was born with. He gained his musical chops the old fashioned way, through deliberate practice and sleepless nights decoding the poorly recorded and scratchy sounds of old "race records." Praise of his 'talent' seemed to erase something Ainslie was proudest of: the hours of hard work he'd put into his passion.

Making the switch from fixed to growth mindset is hard. Growth activities (struggling, sometimes failing) are unpleasant. You have to rewire your brain to let go of comforting labels about yourself, like "naturally gifted."

Dweck's work has shown that mindset is something you have control over. In Chapter five we'll get into the mechanics for building brain apps. In the meantime keep the following in mind:

Chapter Key Points: Hacking your Mindset App

- If fear is a factor in your decision-making, ask yourself, "Realistically, what is the worst thing that can happen?" Often, fear is overblown.

- It's not about where you start on the field of talent. It's your willingness to relentlessly pursue your long-term goal.

- Mistakes—and the lessons we learn from them—are a natural and necessary part of improvement in any field. With the right attitude, challenges make us stronger, not weaker.

- You can only continue to grow if you keep an open mind to change.

- Don't beat yourself up when you mess up. Evaluate what went wrong and determine how to move forward.

- Growth mindset isn't an all-or-nothing proposition; it's a continuing series of decisions.

- In order to increase the strength of your brain's wiring and your growth mindset app, you may need to practice reframing your thoughts many, many times.

- Growth mindset is not a genetic predetermination. It is a point of view, and one that you can teach yourself.

Goal Strategy App

Time to Order the Coffee

Once you recognize the power of framing your experiences with a growth mindset, you're off to the races. Well, not exactly; you need something else. You need a strategy. And that means knowing how to set a goal and use your time wisely to reach it.

We're so busy these days that whenever we even hear 'time management', we tend to shrink back. 'I'd love to reorganize my schedule, but I'm buried right now. Maybe when I get a little more free time.' It's a common line of thought, but consider this.

Imagine you walk into your local coffee shop one bright Monday morning, and when you place your usual order with your favorite barista, she says,

"Actually, we're out."

"Of the Italian dark roast?" you say. "That's okay, then I'll have a—"

"Oh no," she says. "We're out of coffee. All our coffee."

You stare back blankly.

"The thing is, we were swamped last week, and we didn't get a chance to get our order in. We all kept thinking we needed to get it done, but you know how it goes. We had to draw up the menu board and wipe down the display case, and well—one thing led to another. But hey, can I interest you in a croissant?"

Your first thought is that she's kidding, of course. Coffee is a morning basic. It's a necessity; it ushers in your day. Your brain has come to rely on that jolt of java. This can't be happening.

You rush through the door, calculating how much time you'll lose on your way to the nearest Starbucks. The day held so much promise—you were even a few minutes ahead of schedule—but now that's all out the window. All you can think, scurrying back to your car, is, 'That is no way to run a business!'

And yet, how many of us operate this way? The key tasks we know we need to do are precisely the ones we keep putting off, until the mad dash against time begins.

Without a strategy, we just move from crisis to crisis. We spend our day perpetually struggling to stay a step ahead of the problems that keep springing up.

We've become great firefighters, putting out our problem fires, but then we go home exhausted. We decompress in front of the TV or the computer, we go to sleep, and then we wake up and do it all again.

Responding to fires makes sense; often, we don't have a choice. At work, our bosses cause some fires, some are caused by clients or co-workers, and some we accidentally light ourselves. Some fires are like unsuspecting lightning strikes that leave us scrambling for cover.

We justify our firefighter approach because there is danger in just letting them burn. The problem with firefighting is that it's all-consuming. Urgency demands immediate action and a large portion of our energy. It takes away the opportunity to make progress on the things we really care about.

Deep down, you know it's important to move forward in the big picture. Whatever your pursuit—business, leisure, academic, artistic, athletic—progress requires setting and achieving goals. Yet many of us get so caught up with the urgent fires that we let our aspirations slip by. And without staking out a goal and working on it, nothing is going to happen.

Let's leave the urgent fires smoldering a little longer before we address them, but first let's talk about attacking those long-term goals. After all, they're usually the first thing to get kicked to the curb when the smoke starts to rise.

Long-term Goals

Imagine a job where you have total freedom in how you spend your time—no meetings, no supervisors piling on the busywork, and no managers breathing down your neck. Imagine being totally free from company politics but having the ready respect and support of all your co-workers. Imagine spending every day devoting all your energy to a project you believe in so much, you're willing and able to perfect every detail.

In 1993, 27-year-old engineer Ron Avitzur found himself in just such a position. He was happily putting in long hours developing cutting-edge graphing software for Apple, a program that created beautiful 3D renderings of math equations. It was his dream project.

There was just one small problem: Avitzur wasn't technically allowed in the building anymore. The project had been canceled and Avitzur had been let go.[42]

But he wasn't going to let such minor details stop him. "I was frustrated by the wasted effort and so I decided to uncancel my part of the project," wrote Avitzur.

Avitzur enlisted the help of friend and former co-worker, Greg Robbins, whose contract had also just ended. Robbins told his manager that he would now be reporting to Avitzur. His manager bought it, so he was able to keep his security badge. Meanwhile, if Avitzur was asked, he said that he reported to Robbins.

This was enough to let them generally go unquestioned. "I relied on the power of corporate apathy," Avitzur writes. "Since that left no managers in the loop, we had no meetings and could be extremely productive. We worked twelve hours a day, seven days a week."

One day, a confused facilities manager asked why Avitzur's office wasn't on her floor plan. Avitzur cheerfully explained his project had been canceled, and he and Robbins were no longer employees. Furious, she deactivated their badges and banned them from the premises. From then on, every morning, Avitzur and Robbins would sneak into work behind a throng of other employees, and hide in abandoned offices.

It may be worth noting that Apple in 1993 was not quite the visionary tech giant of today, cranking out iPhones and iPads. "It was after Steve Jobs left, and before he returned," Avitzur told us when we interviewed him:

> "There was such a brain drain, there was an evaporative loss of talent. Apple had a salary freeze for a couple of years. One engineer raised this as an issue in a public meeting: 'Look, we're losing valuable talent, because of the salary freeze.' The VP just looked at the engineer and said, 'HR has a thousand resumes. We can hire replacements.' One fellow I knew kept a list of [Apple] engineers on the wall outside his office. As people found jobs and left, they would sign their names. When he left, he wrote his name at the bottom and turned off the lights.

> "In short, it was not an encouraging environment for anyone. I knew one fellow who had been there five years and been through five canceled projects. There were a lot of things that would get started, and when they realized both how difficult it was and how expensive it would be, they would get to a certain point and then it would be canceled and they would start something else. It would be the same cycle all over again."

In a strange way, once Avitzur and Robbins went off the grid, this very issue worked to their advantage. To a building full of stressed out engineers, frustrated by the endless red tape and doomed projects, Avitzur and Robbins became office folk heroes of a sort, and many people were willing to help them—off the record, of course.

Still, it was a tenuous situation.

"There was an amazing amount of silliness in it and a huge amount of risk that it could go horribly, horribly wrong at any moment. I would tell people the whole story, up to wherever we were at. I would spin a yarn about it, asking people for help with what was going on. I was self-aware in the sense that I could step back and look at it and say, 'Well, at least it will make a good story. I hope we're not all arrested.'"

Was Avitzur driven by the adrenaline rush, the edge of danger?

"Oh God no," he says. "That was the most terrible part. I realized how much effort everyone was putting into it. It wouldn't be a funny story if at the end it was, 'yeah, and then we got kicked out of the building, it all got thrown away, and all of these people did all of this work for nothing'."

But that's not what happened. One night, a stranger slipped into Avitzur's "office" at 2 a.m. He was the man who made the master copy of the newest Apple computer prototype, and he was offering to add their program to the master copy—meaning they could sneak their secret 3-D math program into widespread public circulation.

"Once we had a plausible way to ship, Apple became the ideal work environment. Every engineer we knew was willing to help us. We got resources that would never have been available to us had we been on the payroll...Engineers would come to our offices at midnight and practically slip machines under the door. One said, 'Officially, this machine doesn't exist, you didn't get it from me, and I don't know you. Make sure it doesn't leave the building.'"

In October, the software was nearly ready. Avitzur's friends who were legitimately still working for Apple called in their managers for a mystery presentation. Avitzur gave the program a whirl.

The managers were thrilled. It was exactly the flashy new program Apple needed to show off what the new computer could do. The company had thrown all their resources into building a faster, more powerful machine, without providing any software able to prove it.

Why, they asked, had they not been informed about this incredible piece of software? "I explained that I had been sneaking into the building and that the project didn't exist. They laughed, until they realized I was serious. Then they told me, 'Don't repeat this story'."

Luckily, the higher-ups in this project included the son of a math teacher who saw the educational value of the product, and Avitzur and Robbins were "adopted". They still had to sneak into the building every morning—getting new badges would've involved going through Legal, which didn't seem promising given the institutionalized bureaucracy and red-tape—but everyone on the ground level pitched in to make the project a success. After months of hard work, long hours, and help from a huge number of people, Avitzur and Robbins had built a gorgeous, crash-proof piece of software for Apple—despite Apple.

"We wanted to release a Windows version as part of Windows 98," joked Avitzur, "but sadly, Microsoft has effective building security."

What can we learn from the story of Ron Avitzur? It's a triumph of long-term goals. Avitzur and Robbins were able to make the software of their dreams, in part because operating outside the system freed them from the day-to-day urgent fires and busywork.

Still, working for free is probably out of the question for you. Chances are good that you have a boss, and perhaps co-workers and customers, setting urgent fires all around you. You may not have the luxury of going rogue, of becoming the Batman or Batwoman of your office, an untraceable shadow with the power to set and execute your own priorities on your own timetable.

For the rest of us, perhaps the most realistic chance we have of accomplishing something big in our own less-than-ideal setting is a concept called Management by Objective.[43] Peter Drucker first put it forward in his seminal book, *The Practice of Management*.

Drucker originally designed this system for managers who were faced with numerous work projects and needed an efficient way to allocate physical and human resources. This simple template determines whether a given business goal is worth pursuing.

Management Review's George T. Doran summarized the basic tenants of Drucker's theory in the form of a handy mnemonic acronym, **SMART**.[44]

Doran says your goal must be **S**pecific, and progress toward the goal must be **M**easurable. Vagueness and ambiguity are the enemy. The individual tasks must be **A**ssignable to a given person or group, and the overall goal must be doable or **R**ealistic. And finally, the whole endeavor needs to be **T**ime-bound, meaning you need to be able to hit a certain deadline.

It's proven to be a very workable model and is used by businesses all over the world. You can easily borrow this process for your own purposes, well beyond your work life.

From exercise programs to learning a musical instrument to improving your golf game, this template for goal clarification increases your chances of a successful outcome, regardless of your goal. But perhaps equally as important, it saves a lot of time and heartache by sorting out some of the more far-fetched dreams that never really had a chance to succeed.

Successful People

Avitzur described the process of his software project as organic and "ad hoc." When we spoke to him, he was unfamiliar with Drucker's concepts. But looking at his story, it seems in some ways he found himself operating within the framework of SMART all the same.

From the start, he and Robbins had a very **S**pecific goal. As he told us, "We had a very clear vision of what we wanted to do." They wanted a user-friendly program that could create beautiful 3-D renderings of equations.

Avitzur's growth mindset allowed him to not only think this goal was doable, but that through trial and error he would get there. We can see as time went on, this belief continued to drive him forward, even through hours of frustration, even when he wasn't getting paid.

The second principle is that the individual steps toward achieving the goal must be **M**easurable. In other words, there is a verifiable way to demonstrate progress. These are sometimes called gates, certain tests you must pass to make sure you're moving toward success.

For Avitzur, it would be independent verification.

Avitzur wrote, "We needed professional quality assurance (QA), the difficult and time-consuming testing that would show us the design flaws and implementation bugs we couldn't see in our own work. Out of nowhere, two QA guys we had never met approached us—One guy had a Ph.D. in mathematics; the other had previously written mathematical software himself. They were a godsend."

Assignability for Avitzur became a potential issue. The team grew from himself and Robbins to include a whole host of players, from friends doing Avitzur a favor, to allies who believed in their aims, to those QA guys, who were simply very, very bored and looking for a more exciting way to spend their workday. Avitzur even hired a graphic designer out of pocket.

In short, the programmer found himself as a de facto manager, with one key difference: "Nobody had to do anything I said."

In contrast to the stereotype of the antisocial engineer wishing for a tower of solitude, Avitzur attributes much of the program's success to the people around him, and how willing they were to fill in the gaps of what Avitzur and Robbins couldn't do with ease. "It was very much—I asked people to help, people stepped up as they had time and ability."

> "The team was definitely very effective, in that the problems we had, we could find someone to solve those problems very quickly. When you run into something outside of your expertise, it can take you much longer to learn all of that. Working with people who have the expertise you lack, and working on the problem with specializing, is much more effective."

Was Avitzur's software goal **R**ealistic? In hindsight, we know the answer is yes because he and his cohorts were successful, but this phase in goal setting is where the rubber meets the road. Avitzur was willing to bet months of his life on his belief that their software had legs. Were there people at Apple who thought otherwise? Undoubtedly, since the project got cancelled.

When NASA is deciding whether or not to launch a rocket, there is a go or no go aspect of the mission. This is akin to the reality check in SMART strategy. Growth mindset readies you for taking on a new adventure, but it can also tilt toward optimism bias, the belief that the end result won't be as difficult to achieve as it really is.

Someone who has worked on a similar task or goal can give you valuable feedback in what you might expect along the way. It's a worthwhile investment for vetting your idea during this critical reality-check phase.

Which leads us into Management by Objective's Time-bound phase. This phase for Avitzur, and most humans, is one of the most difficult: predicting how much time and effort you'll need to reach your goal.

"When I first asked Greg to help me after the project was canceled, I told him there was about a month left to finish it," Avitzur told us, "and he multiplied that in his head by a factor of two or three, because you always underestimate."

Fortunately, the new Apple computer faced significant unexpected delays of its own, leaving Robbins and Avitzur with six months to iron out every wrinkle of their software. Clearly, even brilliant engineers often don't go into a project able to foresee how long it will take.

Why are most of us of poor judges of time as it relates to tasks and goals? Writer Scott H. Young thinks he knows the reason. In his article "Two Types of Growth",[45] Young describes two different patterns that are generally not considered when we take on a new task or goal.

Logarithmic Growth

Imagine boarding an elevator at ground level in a 30-story building. You effortlessly shoot up the first 15 floors. Then the elevator begins to slow, until finally around the 20th floor, you've ceased to move altogether.

Now to continue the climb, you've got to tackle the last 10 flights of stairs by foot. And much to your chagrin, with each successive floor, the stairs themselves begin to get wider and taller. Making it all the way to the top requires an enormous amount of effort, much more than anticipated.

Sound familiar? Young would suggest maybe you're dealing with one of the following: athletic performance, weight gain or loss, productivity and mastery of a complex skill.

Exponential Growth

Instead of encountering the elevator at ground level, you start out on a series of steps of varied depth and height. It's only after an exhausting climb that, huffing and puffing, you eventually reach the 20th floor. There, you find a shiny new elevator waiting to take you the rest of the way. And as the elevator climbs toward the top, it gets faster and faster.

Young includes the following areas that follow this pattern: technological improvement (e.g. Moore's Law), business growth, and compounding interest.

When we take on a goal, many of us mistakenly believe progress will follow at a steady, predictable rise. This is especially true in business forecasting. Our brains tilt toward this belief so often, that it has a name: planning fallacy.

An often-cited case is the Sydney Opera House in Sydney, Australia. The planners predicted it would be completed in 1963, at a cost of $7 million. In fact, it was finished a decade late—at a cost of $102 million. This constitutes a 1400% cost overrun.[46]

Misidentifying the rate of growth can be hugely problematic. Perhaps it won't be as disastrous as the Sydney Opera House, but when you encounter that exponential climb, you might choose to give up before you even begin. Consider all of those canceled Apple projects of the early nineties, begun and then abandoned once management realized how much work was going to be involved.

Conversely, a logarithmic pattern can give you the illusion of an easy elevator ride straight to the top—until you discover there's a bunch of grueling steps still left ahead of you. It's at this point you take that elevator back down to the lobby and get out.

When it comes to the learning curve for a task or goal, try to understand ahead of time if you're facing a logarithmic or exponential growth pattern. Remember, it's highly unlikely your progress will be constant and consistent. Understanding what might be coming around the curve allows your brain to better guide your inner monologue, so fear and doubt don't sneak in to sabotage your efforts.

It's these obstacles that make fixed mindset so dangerous. At first setback, your inner monologue starts to chant, "This was doomed from the beginning". It's often a self-fulfilling prophecy.

On the other hand, growth mindset reminds you to expect and accept struggle as an essential part of any successful journey.

Your Next Goal

All goals don't have to be as daunting as launching a rocket or developing a cutting-edge piece of software. Imagine you want to get into better shape by honoring Bill Bowerman's jogging legacy. The **SMART** model lets you translate a very nebulous idea into something explicit and actionable. In other words, something you can actually do.

Specific: I will jog two miles every other morning.

Manageable: I need to set aside 30 minutes every other morning in order to jog those two miles.

Assignable: I'll keep my new running shoes by the back door. When the alarm clock rings at 6:30 each morning, I'll lace them up and go.

Realistic: I won't go from 0 to the full distance overnight; I'll start with two blocks and then add an additional block each morning until I've achieved my goal.

Time-oriented: I set deadlines: "By the 15th of next month, I will have worked my way up to running one full mile. By the 15th of the following month, 2 miles."

SMART lets you get specific about tasks to reach your goal, and specificity is key to making an idea a reality. It's tough to achieve a goal without concrete steps to get you there. This book was specifically designed with that in mind.

Cognitive scientists have come to understand there is one more crucial ingredient to successful goal setting: a reward. Working toward a tangible payoff greatly increases the likelihood that you'll stay focused. Whether it's a special dinner, a donation to your favorite charity, or a massage for your sore feet, the key is to treat yourself to something that matters to you.

Specific, **M**anageable, **A**ssignable, **R**ealistic, **T**ime-oriented, and **R**ewarding: **SMART** just got **SMARTR**.

To Do Checklist

To keep the urgent fires at bay, here is, bar none, the easiest, and quickest way to make your schedule work for you. Every morning, make a list of all the things you need to get done that day. Then rank them. Rank the time-sensitive, unavoidable things you need to do with a number 1. For instance, a morning doctor's appointment is a 1 item.

Rank with a 2 the things you'd like to get done. If you want to get a report out, but your boss is on vacation so it's not urgent, that's a 2 item.

Rank with a 3 those items you know you won't get done that day, but you don't want to forget. Writing tasks down frees your brain's working memory and conserves mental energy resources.

The human brain has a very, very small working memory. It can hold only four to seven pieces of information at a time before something is lost. This is why you can run into the grocery store to pick up a couple of items and come outside to find that you can't remember where you parked your car.

For many, this is a mundane occurrence. But a car on average is 22 feet long, five feet wide and five feet high, weighing in at around 4500 pounds. How can our brain lose something that big? The truth is, our working memories fail us on a regular basis, causing us to lose all kinds of things.

If you do just this tiny bit of time management, you can offload those nagging little tasks onto a sheet of paper, a phone app or a computer program. The result? No more forgotten errands, and you free up your working memory to focus on something else more important.

How powerful is this?

In 1927, Charles Lindberg, a.k.a. Lucky Lindy, took off from Long Island on the first successful nonstop Atlantic flight in history.

Lindberg was the first to adopt his own prioritized To-Do checklist for air travel. That list, now used by commercial pilots everywhere, helps guarantee that prior to flight, and during landing, all essential moving parts are properly aligned. Statistics show that, in part thanks to Lindberg, your chances of dying in an airplane crash are something like 1 in 5 million[47] (about the same as finding room for your suitcase in the overhead compartment of today's modern aircraft).

In *The Checklist Manifesto: How to Get Things Done Right*,[48] Atul Gawande cites multiple examples of where the simple idea of employing a to do checklist has seen remarkable results.

In 2001, a critical care specialist at Johns Hopkins Hospital named Peter Pronovost came up with the following checklist for surgery:

- ✓ Wash hands with soap

- ✓ Clean patient's skin with chlorhexidine antiseptic

- ✓ Put sterile drapes over the entire patient

- ✓ Wear a mask, hat, sterile gown, and gloves

   ✓ Put a sterile dressing over the insertion site

After a trial period of just over two years, calculations showed that Pronovost's list prevented 43 infections, 8 deaths, and saved the hospital $2 million.

You might be shocked to discover that a surgeon could need a reminder to wash his or her hands, but neuroscientists know the working memory is remarkably unreliable and prone to errors. We tend to operate like we have unlimited storage capacity. In actuality, there's very little free room left for immediate memory retention in our overtaxed pre-frontal cortexes.

According to Gawande, despite the fact that hospital checklists have repeatedly proven to save lives and money, doctors are reluctant to use them. It makes them feel inadequate, given the implication that they can't remember the simplest of tasks. Still, a little humility seems like a small price to pay for safety.

Planning and Stress

Studies consistently show that prioritizing and planning can reduce your stress level. Even vacations, the very thing most of us believe will leave us feeling refreshed and reinvigorated, can have the opposite effect.

In a February 2014 article for *Harvard Business Review*,[49] Shawn Achor reported on a study in the Netherlands that showed it was common for people to return from vacation feeling no reduction in stress. After all, being in an unfamiliar environment causes all kinds of pressures: reorienting yourself, getting around, and making decisions. It can almost be more taxing than staying home.

Achor looked into this, partnering with researcher Michelle Gielan for their own study. From polling hundreds of travelers, they found that the problem wasn't vacations themselves; 94% of responders reported feeling good after a successful trip. The problem was badly planned vacations.

Advance planning seemed to be a big factor. 90% of the people who reported pleasurable vacations also reported having figured out the details a month or more ahead of time.

You can do yourself a huge favor if you master a modicum of time management. Freeing up your working memory really can make life so much easier for you, whether on the job or on holiday.

Stack Ranking

Want to make a prioritized list even more effective? Once you've put numbers next to all of the items on your list, it's time to stack rank them.1A is your most important, must-do task. For example, if this book was on fire, 1A might be dropping it in the sink and finding a fire extinguisher. Item 2A isn't quite as critical and 3A is a little less urgent.

Stack ranking is key. It's easy to get bogged down on urgent-seeming tasks that aren't, in the big picture, that urgent. We could call this the "squeaky wheel syndrome," as in; the squeaky wheel gets the grease. Assigning a level of value to a more important task helps you guard against wasting precious time on the squeakier ones.

If you can make it to 4A, that's even better, but that's not always going to be the case. Shoot for crossing off your 1A through 3A tasks each day.

This is vital. The truth is, the human attention span is not capable of letting you effectively time manage your entire day, like a conductor directing a well-rehearsed orchestra. That's the myth of the time management industry.

The urge to try to accomplish an unrealistic number of tasks is one of the primary reasons many people fail at time management, and as a result, they don't organize at all.

Status	Rank	Daily Task List
✓	1A	Dentist for root canal
✓	1B	Knock out expense report
	1C	Order new printer cartridge
	2A	Write thank you note to Mom

Time Triage

It's unlikely you'll get to everything on your to-do list. In his book, *The Organized Mind*,[50] Daniel Levitin describes the whole trick to time management in terms of "triage." It's from the French verb *trier*, which means to separate, sort or select. On a battlefield, it's assigning the degree of urgency when treating the wounded.

Since you only have a limited amount of time, you want to be as smart as possible in deciding what you're going to work on and in what order. This means that once you've designated your 1A, 2A, and 3A tasks, you force yourself to work on the tasks in that order. You complete 1A before moving on to 2A, and so on.[51],[52]

It's tempting, once you've created your task list, to pursue the easy stuff. It feels good to cross things off a list. Your brain even rewards you with a little squirt of dopamine, the 'feel good' hormone designed to reinforce behavior. You have to guard against knocking off all the simple stuff and saving the hard tasks for later. That's what turns a small smoldering problem into a raging urgent fire. Adhering to your stack ranking creates a simple and powerful strategy. If you were in the coffee business, a 1A would be to order more coffee beans.

Remember, the items ranked 2 on your list today could become the 1's of tomorrow. Generally, there's no point in stack ranking your 2's and 3's; you likely won't even get that far. But offloading your 2's and 3's on the list helps you avoid overloading your working memory, as well as that unpleasant shot of cortisol when you panic and forget to have cash on hand for the baby sitter.

Time Blocking

In order to tackle the 1-ranked items, you'll need to practice something called time blocking. This requires you to set aside a segment of your day, preferably in the morning when you're still fresh, to specifically work on your prioritized checklist. Time blocking is standard fare for daily prioritization programs.

That said, here is where time blocking holds the secret for tackling your long-term goals as well as your urgent daily tasks. Start with setting aside 15 minutes a day for your long-term goal. Interestingly, most people will tell you how important their long-term goals are, but seldom set aside time to actually work on them. The key is to put your long-term goals on equal footing with the urgent tasks on your checklist.

Status	Rank	Long-term Goal (Boston Marathon)
✓	1A	Research Boston Marathon qualifying times and running programs
✓	1B	Buy new running shoes
✓	2A	Lay out beginning walking course
	2B	Start saving for plane ticket to Boston

Depending on the nature of your long-term goal, you'll discover just how much time you'll need to block in order to make regular progress. Even a mere 15 minutes a day will add up to over 90 hours a year. Ninety hours is not an inconsequential amount of time and will move you one step closer to achieving your desired outcome.

Remember, working **SMART** requires you to stop and re-examine your tendency to bias toward urgency. Make sure you include your intended long-term goal in time planning.

Dealing with Distractions

When you block out your time, turn off your phone, silence your email, and temporarily wall yourself off from outside influence. Distractions have an incredibly negative effect on your ability to stay focused, and, therefore your ability to get things done.

Microsoft did a study on time usage and discovered that every time you succumb to a distraction,[53] your brain takes roughly 10-15 minutes to get back to full focus on your original task. That means if you check your email constantly, you are never employing your brain's full focus power.

What determines how your attention gets doled out? You can thank your brain's insula for that, says Levitin.

Your digestive system breaks food down into a sugary substance, glucose, which fuels both your brain and muscles. Your glucose supply depends on what you've recently ingested and what you've stored from previously digested food. Because your resources can vary from hour to hour, and sufficient glucose to fuel your brain is critical for survival, your brain tends to follow the same energy curve as a laptop computer. When not in heavy use, it powers down.

To conserve energy, your brain automatically slips into what neuroscientists call the 'default mode,' named by researcher Marcus Raichle. This is a natural resting state somewhere between sleep and conscious attentiveness. Most of us call it daydreaming.

Conversely, whenever you do pay attention, your brain is burning your glucose reserves. You literally *pay* for that attention, and the cost is that you can't give that attention to something else. It's a zero sum proposition. The average human's bandwidth for attention is about 120 bits per second. To put that in perspective, Levitin says engaging in a conversation burns about 60 bits.[54]

That's why, when a telephone rings or your email buzzes or a police siren screams or a co-worker interrupts you, your brain can hit attention overload. You've effectively tipped your capacity meter.

And those are just some of the external distractions; add to this your internal ruminations about other problems or concerns, and you have a potent recipe for mistakes, miscommunication and reduced productivity.

In a 2010 paper Levitin, along with Vinod Menon, describes the insula as acting like a mini attention routing switchboard.[55] Essentially, it's a binary switch, absorbing information and instructions from all over the brain and then regulating that input by either connecting your central executive system (the home of reason, speculation, planning, assessment and focus), or instead, plugging into your mind-wandering mode of energy conservation and free association.

The insula biases toward daydreaming—unless it detects possible threats, in the form of changes or details it considers important.

When you're driving along your usual route and you hit road construction, it's the insula that receives the message that the conditions have changed. Levitin says that It's also your insula that immediately detects something is wrong when your friend's voice sounds "odd or different" on the phone.

Or suppose you're in a meeting in mid-conversation with a co-worker and someone across the room mutters your name. There's a good chance you'll immediately pick up on that. How is that possible? Your insula is tuned to what you care about. In this case, your name acts like a trigger for your attention.

When you're shopping for new cars, and you suddenly start seeing certain makes and models on the road over and over again, Levitin says that's your insula at work. There aren't suddenly more of those cars, but rather your neural circuitry has put itself on the lookout.

Levitin calls this the "Where's Waldo effect", after the popular children's game of trying to find that iconic character wandering in his red and white stripes amongst a sea of similarly colored surroundings.

Tuning In

When your brain identifies change or things of import, it begins retuning part of its circuitry for better recognition. This is partially what we mean when we talk about neuroplasticity, or the brain's ability to reorganize itself based on what you need it to do.

This retuning of attention for specific purpose explains how quarterbacks are able to spot receivers amidst shifting defenses, how pilots can find the right runways, and how you can single out your own child from an ocean of kids in a crowded auditorium. Attentive retuning aids you across a spectrum of important activities.

The good news is that your insula is always on high alert. Unfortunately, when you're in work mode, the factors it watches for can often fall under the heading of distraction. Your brain has been fine-tuned to stop what you're doing and pay attention to the sound of your boss's voice, the tone of your phone chime, a song by your favorite band that just came on the radio, and a galaxy of other external signals.

The insula's high alert default is your brain still running programming from an evolutionary past hardwired to see distractions as potential signs of danger. Since there is no brain upgrade coming any day soon, you need to eliminate as many of those signals as possible.

This has nothing to do with intelligence or physical stamina. In the last hundred years, our technology has quickly overtaken our brain's processing capacity. Nobody's brain has the bandwidth to absorb and hold the 300 Exabytes of information we encounter in our everyday lives. (Levitin points out that 300 Exabytes, spread out as readable text, would cover every inch of Massachusetts and Connecticut combined.) We are forced to deal with what engineers refer to as "inherent design limitation."

Ron Avitzur, the Apple programmer we met earlier, found himself operating incognito, escaping the drama of standard office life. Limited distractions allowed him to uni-focus on his task.

The writer William Faulkner minimized disturbances when he was writing by removing the doorknob to his study so no one could enter. He reportedly carried it around with him in his pocket.[56] Probably a bit extreme, but it does a nice job of driving home the point.

We want the fastest, most powerful machines and software we can get our hands on. This urge fueled early sales for BlackBerry—and, as noted, also led to their demise. People have been known to wait outside all night long, enduring inclement weather and long lines to secure the latest model of a smartphone.[57]

And yet the most powerful machine, the one that calls the shots—the brain, the mechanism designed to keep us alive—we often neglect. Until fairly recently, many of us haven't been concerned with optimizing the brain as a tool to the same extent that we are with our computers and devices.

This was partially due to a lack of understanding about the rudiments of brain function and subsequent behavioral implications, and partly because it's awfully easy to take our brain for granted. We don't notice the quiet hum of our own cognition until there's a problem.

The Myth of Multitasking

When it comes to problems, multitasking can be deadly. In the tragic story of Eastern Airline Flight 401, Pilot Captain Robert Loft was making what appeared to be a routine landing at Miami International Airport. In the final stages of his approach, Loft deployed the plane's landing gear and noticed the indicator light that registers 'wheels down' had failed to activate.

He put the plane into a holding pattern at 2000 feet to investigate. Soon the first officer, the flight engineer, and a Boeing mechanic who happened to be sitting up front in the jump seat were all trying to discern the problem. By the time Loft realized nobody was actually flying the plane, he only had time to shout "Hey, what's happening here!" before the plane smashed into the Florida Everglades, killing everyone on board.

Tragically, the problem turned out to be a burned out bulb on the cockpit control panel.[58] The FAA determined there was nothing wrong with the landing gear. Something as simple as an activation light had hijacked four crucial people's attention systems, leaving the controls unmanned. A burned out bulb resulted in the death of 99 people.

As safe as it is to fly, a large number of the accidents that do occur are attributed to this kind of human error. Over an 11-year period in the late eighties and early nineties, the US Air Force lost 190 lives, as well as 93 aircraft for a total of $1.7 billion in financial damages.[59] In over half the cases, the cause was later determined to be "task saturation", more commonly known as multitasking.

The brain, amazing as it is, still does not have the power to truly perform multiple complex actions at once. What we take for multitasking is actually the brain's insula flipping back and forth between tasks, working so fluidly, it gives us the illusion of simultaneousness. Neuroscientists call this "attention switching."

Attention switching between conscious activities burns glucose much faster than focusing on a single task. So, attempting multiple mental activities, like text messaging, while conversing, while simultaneously checking your car's speedometer is an energy drain.

Multitasking shouldn't be confused with biomechanical activity like your ability to walk and chew gum at the same time, because neither of those actions requires tremendous conscious thought.

Trying to do two complex things at once forces your brain to run sub-optimally, tires you out faster, and predisposes you toward mistakes. And once you get tired and run out of gas, it's easy to find yourself stuck in *procrastination roundabout*.

The Trap of Future-Self

Our brains have what scientists believe to be a very rare quality in the animal kingdom: the ability to project into the future. This allows us to save mental and physical resources by simulating ideas in our brain's basal ganglia, so we don't have to actually try everything out in real time.

This is a pretty cool adaptation. Think about flight training. Nobody wants to jump in a jumbo jet to find out that this is a pilot's debut at the controls. Instead, flight students use simulators. Flight simulation has undoubtedly saved millions of dollars and an untold number of lives.

In much the same way, your brain's ability to simulate future events allows you to work through all sorts of scenarios that might otherwise prove to be time consuming, embarrassing, or even deadly. But there's a downside to mental simulation: it also lets us imagine ourselves doing a given task at some later date.

Why do something now if you can save it for your Future-Self? Especially when you can dream up a Future-Self much better equipped for the job—well-rested, more engaged, and ready to take on the world. Future-Self is your mental super hero, undaunted by piles of work, expense reports or your workout program. By offloading a task to the person you might be later, not only can you skip a task right now, but you can fill that time with something Present-you finds much more enjoyable.

In a way, procrastination is a genius move: it's not laziness, it's delegating to a *you* that is always just about to exist. "I'm too tired to start that job right now, but you know who's going to have energy to burn later on? Future-Self." Future-Self is a bit of psychological jujitsu.

This is how people with fixed mindset can fool themselves into believing they're more growth oriented than they really are. Delegating to Future-Self allows you to stay safely planted in your comfort zone. It makes you feel like you haven't crossed any of your options off the list. Future-Self lets you stay flexible, be safe and conserve your energy.

There's only one problem: Future-Self is never going to show up. It's like chasing the horizon: by definition, Future-Self is always just out of reach, constantly being replaced with Present-Self as time marches inexorably forward.

Our brains are constantly tricking us into thinking we're going to work on a task or goal down the road. And these forecasts are frequently wrong. As Scott H. Young points out, when we try to predict our progress, we tend to imagine constant improvement at a steady rate: a forty-five degree angle pushing away from the X and Y axis like a rocket ship.

In reality when the actual learning curve doesn't match up with our imagined progress, we get discouraged and often, we abandon our goal.

Sometimes rather than admit we'll probably never come back to a task, we instead delegate it to a newly conjured Future-Self, who will later take it up at some elusive unspecified time. This way we don't have to acknowledge we've come up short.

Marketing people figured this out a long time ago when they developed rebate programs. It is estimated that between 40-60% of people never get around to collecting those advertised rebates. Consumeraffairs.com says this amounts to $500 million a year in consumer loss.[60]

Out of Time?

At this point, you might be agreeing that having a growth mindset and **SMARTR** goals would be great, if only you had the time to work on them. But you're so busy with _____, there's just no way you can think about tackling a long-term goal, as much as you'd love to. It's just the reality of your life right now. You wish it were different.

If you found yourself nodding your head or agreeing with the last paragraph, there is a reason for this. Many of us have wired our brains for the "I'm so busy" response, and we sincerely believe it. Most of us can back up our claim with a long list of all the things that eat up our day and leave us exhausted.

Wouldn't it be nice if we had an extra hour in each day? Imagine how much more you could get done with another 60 minutes of free time. You'd sign up for that, right?

Unfortunately, the earth's rotations never got the memo, and we're stuck with our 24-hour system—and once you pull eight hours of sleep out of that, you've got just 16 waking hours left to order the metaphorical coffee.

Or do you?

A study by the Council for Research Excellence suggests that on an average day, adults are exposed to 8.5 hours of screens—TV, computers, smartphones, and so on.[61] Now most of us are committed to a certain amount of screen time; we can't exactly tell our bosses we're on strike from answering work emails or looking at spreadsheets. But even factoring out our work-related usage, it's still likely that each of us is spending about three hours of our precious downtime on screen-related activity.

When you think about how nice it would be to find some extra time in the day, you don't generally think about this time as negotiable. After all, it's what you do for fun. It's part of your me-time.

If we were to analyze our three hours of screen related downtime, we would see that indeed, some of it is enjoyable—catching your favorite show, for instance, or trading messages with your best friend. These experiences tend to be what tech consultant Linda Stone would define as a purposeful but relaxed presence: watching something with clear intention.[62]

However, of that three hours, there's probably a sizeable chunk in which you're a little spaced out, not fully "online", so to speak: flipping through channels, reflexively refreshing your email, idly scrolling through your Facebook feed.

"The Internet is not addictive in the same way as pharmacological substances are," says Tom Stafford, a cognitive scientist at the UK's University of Sheffield. "But it's compulsive; it's compelling; it's distracting."[63] And distraction often takes the form of 'surfing,' killing time until something more exciting comes along. Frequently, it doesn't.

Maybe not surprisingly, studies show that little pleasure or enjoyment registers in the brain during this kind of activity. It really is, as far as the brain is concerned, much more of a holding pattern. You're not riding a roller coaster of fun, you're waiting in line for what will hopefully be the next ride—and it's about as thrilling as waiting in a line ever is.

Psychologist Mihaly Csikszentmihalyi points out that "hobbies are about two and half times more likely to produce a state of heightened enjoyment than TV does, and active games and sports about three times more."[64]

Csikszentmihalyi cites a wide-scale German study that "passive leisure

[I think it's fair to include surfing in this category] becomes a problem when a person uses it as the principle—or only—strategy to fill up free time. As these patterns turn into habits, they begin to have definite effects on the quality of life as a whole."

So if your "recreational" screen use is roughly 50% purposeful and 50% passive or surfing, that means that every day, 1.5 hours go down the drain.

Furthermore, we know that in skill acquisition mode, the adult brain can basically hang on for about 15 productive minutes before attention starts to wane.[65] So those daily 1.5 hours represent six fifteen-minute segments of opportunity you could be time blocking toward the goal of your dreams.

Just stealing back two of those 15-minute segments a day for one year is over 180 hours. That's more than a month's worth of time—enough time to make significant progress toward learning a foreign language, a musical instrument, or improving your chess or golf game. Not to mention kicking your gardening or gourmet cooking skills up several notches. You'd have time to meditate, or exercise or bone up on the latest sales technique.

And you don't have to knock off your screen time. Just be more purposeful in your viewing, and stop when you find yourself slipping into surfing mode.

Nobody likes to stand in line. The Disney Corporation understood this early on. They created the snaky line illusion, which tricks the brain into believing that the serpentine queue of humans waiting for a ride at their theme park isn't as long as it really is. Screen surfing for many of us has become an accepted practice. In the world that Walt Disney created, standing in line is a given. In your own kingdom, it's up to you.

Have you ever gotten stuck clicking from YouTube video to YouTube video, and then before you know it, your Saturday afternoon is gone? Or you sit down to watch your favorite show, and then wind up watching whatever comes on after, even if you don't have strong feelings about it, just because it's there.

We're so predictable that advertisers can rely on us to behave exactly in this manner. They build their ads around our habits. They, too, rely on Future-Self. It's Future-Self that will be more attractive with the right car, aftershave, or lip-gloss. But Future-Self isn't just reserved for your leisure life. This character is also a mainstay at work.

Here again, the ability to project into the future allows us to escape in the middle of a meeting and do mental walkabouts in the land of the Almost-There. I suspect our collective Future-Selves traverse an awful lot of mental landscape during endless PowerPoint presentations.

Now, I'm not saying you should get up and wander out of a meeting or toss your laptop into a pond. Examine your TV routine and promise yourself that when your favorite show ends, you're not going to get sucked into a series you don't even like. Turn off the TV or computer. How many YouTube cat videos do you really want to watch?

Practice the piano. Do a little Tai Chi. Go out and take a short walk. You have at least six of these increments available each day. Do something productive with a couple of them. A few of those 15-minute increments can help fuel your long-term goal.

Without some self-examination, you're primed to activate the "just keep watching TV" app or the "one more YouTube video" app. This is why so many of our habits have crept into our lives without any intention. Your unconscious simply delivers the programming you've wired through repetitive action over time. The basal ganglia makes no judgment call as to whether the brain app you use for screen watching—or any brain app, for that matter—is good or bad.

Your brain's future projection abilities also mean that when you think about doing something more productive in the moment, you tend to project a worst case scenario, imagining the maximum amount of effort it could take. The brain's energy conservation tendency biases you against starting something new. This is part of why Future-Self is so pervasive, with its promise of unlimited energy available down the road.

However, forcing yourself away from a screen and taking a walk, for example, will often leave you feeling better. In a competition for energy resources, you have to be vigilant to guard against your pre-existing penchant for stasis, keeping you glued to your chair or couch.

By deciding on a relatively small increment of time, like a five-minute walk, the activity, and the necessary energy cost, feels far less daunting. Once you're up and walking, the idea of continuing after five minutes is relatively easy. In this way, you can override your unconscious brain's energy conservation strategy and break the reliance on Future-Self. This can serve as the gateway to working toward a long-term goal. (More on this in Chapter Five.)

The Stacker Syndrome

Some of us have an office that's buried in paper. I'm not necessarily talking about tunnels of floor-to-ceiling stacks of yellowed newspapers and magazines, with a dozen cats lounging about. (That's a much more serious problem.) I'm talking about the stacks of paper that are inhabiting a good portion of your horizontal office surfaces.

If this sounds familiar, keep reading because you are what we affectionately call a Stacker. If that's not you, skip this section and go to the next one.

Okay, if you're still here, the first thing to examine is the Stacker's credo, which goes something like this: "It might not look like I'm organized, but I know where everything is in these piles." I'm sorry, but in truth, this is shorthand for, "I actually have no idea what's in these piles."

The Stacker employs an interesting combination of Future-Self and Present-Self. It starts when Present-Self has the realization that an important piece of paperwork (a bill, an invitation, a memo, an order, a letter, some reference material, a correspondence) has shown up in the office.

Stacker protocol is:

1. Stare at the new piece of paper for a few seconds—or an indeterminate period of time, depending on your coffee intake for the day

2. Acknowledge its importance. (Often, this is done out loud: "I better keep track of this.")

3. Set it down somewhere within your field of vision to make it easier for Future-Self to locate later.

4. Repeat above, over and over again.

This strategy might make sense if your office was the size of a football field and you had some rapid means of flying above it all to locate the relevant document. Unfortunately, your office, cubicle, or kitchen table probably has limited capacity and so, for space conservation, stacking ensues.

Now comes the challenge: you have to remember what you've put in each stack. I've often thought this has the making of a reality show: *Ultimate Stacker*. Cue suspenseful music.

In reality, stacking is a tremendous tax on the memory. As we've noted, working memory can only store four to seven items at a time. Stacking gives the illusion of organization, relying instead on the mythical Future-Self to sort it all out later.

Breaking out of Stacker syndrome requires the following steps.

Pick up any piece of paper in your office and ask yourself, "Where does this belong?" Your options include:

1. Putting it in a labeled folder in a file cabinet

2. Relegating it to a recycling bin

3. Prioritizing it on your task list

4. Performing an action that renders that particular piece of paper useless (Like paying a bill, scanning the paper into your computer, and so on.)

You only have four options available; notice there is no option of stacking.

This is important because if you are a stacker, you'll discover that eventually you will be forced to do one of those four options anyway, largely because, as we've come to understand, Future-Self never shows up.

The inescapable fact is: if your Present-Self chooses to stack, you'll still be handling that paper again later, with only the same options available.

Rather than simply delegating the matter for Future-Self, thereby effectively doing nothing, Present-Self has to go through the act of stacking first. It's procrastination with a dose of work added to it. For a hardcore procrastinator (It's estimated that when it comes to taxes, about 24% of us fit the bill.[66]), that's a pretty bitter pill to swallow.

It's not uncommon for Stackers to get a burst of energy, and in an effort to purge their office of paperwork and reclaim horizontal work surfaces, attack their stacks. Like Sherman marching through Georgia, a wake of destruction follows, except here the losses take the form of reams of paper stuffed into the recycling bin.

The next time you encounter a Stacker fresh and smug from victory over a recently purged office, ask them, "Hey, what happened to the 'but I know where everything is in this stack?'" He or she will smile slyly, somehow forgetting that in the same way "a journey begins with the first step," a new stack begins with a single sheet of paper.

Perpetual Inboxer

If you are a person whose inbox is more celestial black hole than email receiving station, a.k.a. "Perpetual Inboxer," it should be noted that email can be sorted in much the same way as above. In this case, your four options are:

1. Delete

2. Prioritize

3. File

4. Act on the email (As in, answer it, or perform some necessary task)

Growing out of these two procrastination modes is a pretty big identity change on your part. You're going to need a boost in motivation. In the next chapter, we'll learn how to ratchet up the willpower effect.

Chapter Key Points: Hacking your Strategy App

• Plan for your long-term goal with the **SMARTR** program. Make your goal specific, measurable, actionable, realistic, time oriented, and reward based.

• Try to anticipate whether your progress will be logarithmic (a fast start which then slows down) or exponential (a slow start which then speeds up). Manage accordingly.

• When setting a goal, try not to introduce more than 10% change in your daily activities per week. Too much change creates stress.

• Prioritization, by necessity, is a two-fold strategy. Operate in triage mode and stack-rank tasks based on urgency, but always leave room for working on your long-term goal.

• Time block and manage your activity in highly focused fifteen-minute increments.

• Distractions are your enemy and there is no such thing as multi-tasking. Uni-tasking will help conserve your precious glucose.

• Recognize when you're delegating to Future-Self and instead hand the reins back to Present-Self.

Chapter 3
Willpower App

Maybe you've managed to evade the allure of Future-Self for the time being and temporarily find your way out of *procrastination roundabout*. But how do you stay out? How do you carry through on your long-term goals? For that, you'll need to build your willpower app.

No discussion about willpower would be complete if we didn't include one of neuroscience's most famous mysteries: the case of railroad foreman Phineas Gage.

Imagine a beautiful Vermont day on September 13, 1848. It's about 4:30 in the afternoon. James K. Polk, 'Young Hickory', is the current president and the Civil War hasn't yet ripped the country apart. The hottest thing in modern technology? The steam-powered locomotive.

He's had his 13-pound tamping rod engraved with his initials and shaped to his exact specifications by a blacksmith, honed to a sleek taper and carefully polished smooth.[67] Looking at it, you might never guess its purpose.

The dinner hour is fast approaching and an exhausted Gage leans over a newly drilled hole, takes up his custom-made tamping rod and gives the explosive charge one last push.

But something goes horribly wrong. The charge explodes on contact. In a split second, Gage's iron rod blows through his left cheekbone, shooting upward past his left eye socket, tearing through his prefrontal cortex and out the top of his skull. The rod lands with a thud some twenty-five yards away, sticking neatly upright in the ground.

Gage is knocked violently backwards. Splattered in blood and bone fragment, somehow he remains conscious despite having lost what will later be referred to as "half a teacup's worth of brain matter."[68]

What happened next? According to Sam Kean in "Phineas Gage: Neuroscience's Most Famous Patient",[69] Gage staggered to his feet, where he was helped to a wooden oxcart and transported back to his hotel room in the nearby town of Cavendish.

When the doctor finally arrived, Gage was sitting on the front porch of the hotel to greet him. "Here's business enough for you," he allegedly said. His condition was worse than the jaunty greeting implied; for weeks, he suffered fevers and infection, his health was touch and go. But Gage didn't die in 1848. He recovered and went on another twelve years.

Here is where the story gets fuzzy.

With the loss of a good portion of his prefrontal cortex, Gage is reported to have turned into a raving maniac, driven by desire, prone to fights and fits of obscenity. The tamping iron had robbed him of more than just a "half teacupful of brain matter"; it had taken his self-control, his willpower.

The problem with this account is that we have no solid evidence to back it up. All we really know is that Gage went on to make a living as a horse coach driver—a job that requires a certain amount of manual dexterity and restraint—and died in 1860.

Reports make it clear his personality was forever altered, but just how much? This is one of the most discussed neuroscience stories in history, because, beyond the shock value, it purports to tell us a few key things about the nature of the prefrontal cortex, the processing center in your skull just above your eyebrows. The prefrontal cortex does a lot of cool stuff. It is, among other things, home to your willpower.

Neuroscientists sometimes say we have one brain and two minds. What does that mean? Psychologist Daniel Kahneman explains that we essentially have two different systems for making decisions.[70] He refers to them simply as System 1 and System 2. It's important to understand that these are not actual physical locales of the brain—the two systems operate on overlapping real estate—but, rather, are simply a template for understanding collections of behavior.

System 1 is automatic, pattern-driven, and reflexive, constantly reading your surroundings to quickly generate short-term predictions and spur immediate action. It relies on emotions as a gauging mechanism. It's also the system that sounds the 'lion alarm' that kicks you into fear mode we discussed in Chapter One.

System 2, by contrast, is analytical and requires conscious effort, weighing pros and cons, allowing you to focus on boring tasks, search your memory to identify something unusual, and monitor the appropriateness of your behavior. Part of its job is to exercise willpower over System 1. You can think of it as the rational mind if you'd like, although it has shortcomings of its own and can be lazy about intervening on System 1's impulsive shenanigans.

As Kahneman points out, when we're hungry and tired, our focus and personal willpower diminish at a startling rate. As the heavyweight fighter Mike Tyson once reportedly said, "Everyone has a plan, until they get punched in the mouth."[71]

In Tyson's case, the insight was probably quite literal. Taken in a broader context, it describes the brain's limited ability to stay on task when confronted with problems, temptations, trauma—or a deprivation of food, sleep or energy.

Reflexive System 1 can get a bad rap even though it's our default system, evolved to err on the side of caution in order to keep us alive. Still, it does get us into trouble sometimes. It tends to bias, weighing some pieces of information over others and it loves shortcuts. Flaws in our System 1 thinking allow us to be fooled by optical illusions.

System 1 also has a huge bias toward noticing and avoiding danger, which keeps you safe, but also generates plenty of false alarms and irrational fears. (System 1 reacts emotionally to even seeing the word 'crime'.) It pushes you to unexamined conclusions.

Reflexive System 1's emotional, instinctive processing causes us to reach for the glazed donut, and it is analytical System 2 that reminds us that we vowed to eat healthier. The winner depends on a lot of factors.

We need willpower to resist the impulse from the emotional mind. Our ability to demonstrate willpower is the best predictor of success that we have. It's more important than intelligence, skill, or wisdom. This is true in nearly any area of life across the board: careers, jobs, relationships, and even general happiness.[72]

Much has been written describing the personal aftermath of Gage's accident. The most often told story is that the once-affable, upstanding citizen was scarcely recognizable, prone to outbursts of temper due to his loss of self-control. Kahneman might describe this theory as a tale of Gage's analytical System 2 mode failing to override his more primitive System 1 reactions.

It's unlikely we'll ever know what Phineas Gage's willpower was like after his bizarre accident. Still, this case demonstrates the brain's remarkable resiliency. After all, despite suffering definite brain damage, Gage was able to live some kind of functional life for another dozen years.

Stanford University professor Kelly McGonigal says, that most of us from time to time experience a little of what neuroscientists euphemistically call "temporary brain damage,"[73] although not on the level of Phineas Gage's cranial piercing. It happens whenever we over-imbibe alcohol, or deprive ourselves of sleep, or find ourselves on the verge of temptation. All of these situations divert energy from the prefrontal cortex and reduce our level of executive control; that is, they reduce our willpower.

But willpower is more complicated than just a simple wrestling match between your reflexive and your analytical decision-making systems. When it comes to maximizing willpower, there are three things you'll need to keep in a near-perfect balance: exercise, nutrition and sleep.[74]

Willpower Triad

We've all heard that exercise, nutrition and sleep are important. It's important to understand these three factors are intricately related, and influence each other to the point of inseparability. They're the foundation of our willpower, not as separate islands but as a holistic ecosystem.

When you get that system in good working order, you have the internal reserves to harness and exert willpower. And no matter what goal you're trying to reach, you need to demonstrate at least some willpower. The problem is that if you ignore one aspect of the exercise-nutrition-sleep triad, the whole system goes out of whack.

For instance, if you aren't well rested, it's going to be almost impossible to summon the energy to work out. When you're tired, you have less energy to devote to food preparation, which allows fast food and low quality meals to creep into your eating routine. Poor nutrition interferes with your sleep patterns, and the cycle continues.

If you can maintain a good balance, at the end of the workday, you should still have some energy left to pursue your long-term goal. If you find yourself completely out of gas, it's time to look under the hood and check your exercise-nutrition-sleep triad.

Exercise

Many Americans frame exercise as the antidote for overeating, a necessary evil. Why do we exercise? Because we don't want to be overweight; we want to feel and be seen as attractive.

Some of us view exercise almost as a punishment: if we splurge on dessert, our shame nags us to put in some extra time on the treadmill. On the other hand, if we've had a stressful day, or a stressful series of days, we let ourselves off the hook and indulge in a little ice cream—until the next time we start to feel bad about our bodies and the cycle continues. Entire industries thrive on offering such temptations, and a fitness empire thrives on our guilt at giving in.

Sticking to an exercise regimen means you likely will start dropping pounds at some point, but that process tends to be very gradual. If your only focus is shrinking the number on the scale, you may get discouraged after a few weeks, or you'll push yourself so hard that you'll risk burning out. Remember the elevator ride in Chapter Two? It's the exponential vs. logarithmic pattern writ large.

There is a better way to reframe your concept of exercise: don't do it for weight loss, or shame. Take those two entirely out of the equation. Instead, exercise because it's one of the best things you can do for your brain function, and to increase your likelihood of hitting your long-term goal. Modern science has invented no brain booster as powerful as exercise. No supplement or memory game can match its benefits—and it's free.

The Science Behind Exercise

Exactly what does exercise do for the brain on a cellular level? For one thing, prolonged physical activity drives more oxygen and blood flow to your brain, which boosts your brainpower. The increase can be up to 20%. In addition, when you regularly raise your heart rate, you produce BDNF: brain-derived neurotropic factor. This substance has been called "Miracle Gro for the brain."[75] It strengthens your signaling power, increases serotonin production (important for learning and self-esteem), encourages neuron growth, and protects existing neurons.

In a German study two groups walked on treadmills. The first group supplemented their 45-minute exercise routine with two separate 3-minute intervals of intense sprinting. The control group didn't include the sprinting intervals. In subsequent memory tests, the sprinting group showed a 20% increase in word acquisition over the non-sprinters. Not surprisingly, the sprinters showed increased levels of BDNF, while the non-sprinters showed no difference.[76]

Other studies support[77] the German findings: added bursts of brief, high-intensity exercise have a profound effect on mental performance. (For an extra BDNF boost, consider consuming foods spiced heavily with turmeric, present in many Indian curries. Still, exercise is by far your best bet.)

Neuroscientist John Medina recommends that we exercise 5 days a week for 30 minutes a day at an aerobic level. Among other things, he says, you'll reduce your chances of Alzheimer's disease by 60%.[79]

Memory improvement not convincing you? How about kicking your mood up a notch? The neurotropic factor released during exercise appears to increase the output of dopamine and serotonin. Both are associated with well-being and enhanced mood.

Physical activity temporarily depletes your store of glucose,[80] the fuel derived from what you've eaten. The brain overcompensates by storing additional glucose in its cell walls. You wind up with reserve glucose, which translates into more processing power, more memory, and better learning capabilities.

What kind of exercise works best? Ideally, you want a combination of aerobic, interval and strength training. A 2016 Finnish study showed aerobic exercise, like running, increases adult hippocampal neurogenesis (AHN). The study indicated that running generated more neuron growth than interval exercise or strength training. Although the study involved rodents, scientists believe a regular morning jog would connote the same benefit for humans. But although running does a great job promoting neuron growth, the study suggests that strength and interval training have other brain benefits too.[81]

Strength training has a powerful effect on memory. Regular strength training for twenty minutes two to three times a week can boost your long-term memory by 10%.[82]

If you're not chasing that muscle-bound, bodybuilder look, don't worry. It takes a tremendous amount of deliberate effort to put on that much visible muscle.

Some clinical studies suggest that exertion to the point of sweating, three days a week, sustained long enough to burn about 350 calories (about 25 minutes on a treadmill or bike for most people), has the same positive impact on mental depression as antidepressants.[83] I'm not saying you should throw out your medications; mental health is a complicated issue. But if you're looking for some way to improve your overall mood, you'll do yourself a huge favor if you start exercising.

It's believed the human body was designed to walk about 10 or 12 miles a day. People evolved to cover large distances prior to the domestication of horses. The more you walk, the better your brain functions. And keep in mind we're not necessarily talking about power walking.[84]

You'll improve your brain's performance by walking at a rate of two to three miles an hour. That leisurely stroll gets you the perfect amount of oxygen and circulation to think at the top of your game.[85]

Steve Jobs believed in walking. When people came to see him, instead of meeting in his office, he'd often take them for walks around Palo Alto.[86]

At night, after dinner, go out for a short walk. If you have a significant other, pet or both, bring them along. A daily walk is a huge treat for your brain. But it gets even better: a walk of 30 minutes after mealtime cuts your insulin level in half.[87] Insulin regulates whether your body is going to use calories you've ingested or store them.

Walking after dinner means it is far less likely you'll be storing those calories in the form of body fat and you'll be optimizing brain function at the same time.

Bottom line, the exercise that is the most valuable to you is the one you'll actually do regularly. With that in mind, research indicates that a combination of walking, strength training, interval training and aerobic activity is the best way to hedge your bets, improve brain function and increase a key component of your willpower.

Nutrition

There is a lot of conflicting information out there about what you should be eating. For brain health, the standout approach, the one with far and away the most research behind it, is the Mediterranean Diet.

Researchers in Spain found that people who followed this program for six years performed noticeably better on cognitive tests than people on a low-fat diet.[88]

Even though "diet" is in the name, the objective is not strictly about cutting calories or counting points. Essentially, the goal is to eat a lot of non-starchy vegetables, lean protein and the right fats. Fat intake is critical for a healthy brain. Between 50- 60% of the brain is made of fat, especially myelin, which insulates neural connections and boosts signal strength. As neurologist and co-author of *The Better Brain Book,* David Perlmutter, says, "Too much fat is better than too little, and nothing is worse for the brain than a fat-free diet."[89]

This isn't to say that you should load up on pork rinds. Diets high in saturated fat (red meat, whole milk, and butter) encourage the risk of brain dysfunction. For example, saturated fat consumption can increase the onset of type 2 diabetes, which can lead to stroke and dementia through blood vessel and cellular damage.[90]

You should also avoid trans fats—the kind found in margarine, fried foods, and many processed foods. The good news is that as of the writing of this book, trans fats are beginning to be phased out of the American diet. And none too soon. A study showed that men 45 and younger who consumed trans fats performed significantly worse on cognitive tests than those who didn't. This has long term implications as they age.[91]

So how do you get the right fats? Make sure you're eating olive oil, nuts, avocados, and cold-water, fatty fish. The optimal brain fat is that famed omega-3 you've heard so much about, especially a type of omega-3 called DHA. Good sources include walnuts, flaxseed, seaweed, sardines, herring, and albacore tuna.

For brain optimization, you also want to minimize your sugar intake as much as possible. In a 2012 UCLA study, rats fed water sweetened with fructose performed significantly worse in mazes. They also showed resistance to insulin, which is what causes type 2-diabetes. This is bad news in more ways than one. In addition to regulating blood sugar, insulin is involved with certain signaling in the brain. So, if strokes and dementia weren't scary enough, a resistance to insulin may reduce learning and memory functions.[92]

Rats fed DHA (the good fat) perform much better than their counterparts. In fact, research suggests that to some extent, DHA can counter damage done by excessive sugar intake.

"Well, it's a good thing I don't make a habit of sipping on fructose-water," you say. Except if you're drinking soda pop, that's literally what you're doing. Fructose is a form of sugar found in both cane sugar and corn syrup, and a key ingredient in soft drinks.

The Brain on Soft Drinks

When we think of addiction, we might think of the heroin junkie lying on a dirty mattress with a needle in his arm, or a sweaty rock star snorting a line of coke right after some mega concert.

We are probably less likely to think about ourselves. But when it comes to hijacking the brain's reward system, many of us are not entirely clean.

You might not realize it but drinking one of America's most cherished drinks, a 12-ounce can of soda, has ominous repercussions.[93] A single soft drink packs about ten teaspoons of sugar per can. Stack this up against the American Heart Association's daily sugar guidelines: nine teaspoons for men and just six for women.[94]

One can of soda pop means you've consumed an entire day's worth of sugar.

What's the net effect?

About twenty minutes after you chugged that can of soda, a blood-sugar spike overwhelms your liver's processing capacity. What your liver can't process is converted into fat.

At the thirty-minute mark, the soda's caffeine has kicked in, dilating your pupils and driving up your blood pressure. The increase in blood pressure sends a signal to your beleaguered liver to release even more sugar into your bloodstream.

Forty minutes after the soda washed over your lips, the levels of dopamine in your brain's nucleus accumbens (pleasure center) go wild. Since dopamine is a feel-good chemical, designed to reward beneficial behavior, the result is a kind of 'high'. As with heroin, your internal chemicals have no way of knowing they've been manipulated into rewarding you for something that is actively hurting you.

Fifty minutes later you've suddenly got the urge to urinate, thanks to the diuretic quality of the caffeine. You're not just losing fluids, though; soda contains phosphoric acid, which binds to calcium, magnesium, and zinc. So "you'll soon be flushing those vital nutrients down the toilet."[95]

One hour after you knocked back the soda, you sugar crash big-time, and then begin to go into withdrawal. You're irritable, your energy levels flag, and you're thirsty. Your solution? Another can of soda.

If you're watching calories and your drink of choice is diet soda, does the above still apply? Yes. In fact, it's worse. Artificial sweeteners in diet soda have an even greater impact on your brain than regular soda.

The Journal of American Geriatrics Society did a nine-year study involving 749 adults over the age of 65. Regular diet soda drinkers gained almost three times more belly fat than non-diet soda drinkers. Storing additional fat, particularly around the belly area, is associated with a higher risk of both heart disease and type 2-diabetes.[96]

How could we see these results from diet soda, a beverage designed, at least in theory, to be healthier? Sharron Fowler, of the University of Texas Health Science Center at San Antonio suggests that there might be two reasons.[97]

First, it might be that people who regularly consume diet soda trick themselves into believing they can eat more junk food as a result of the savings they've accumulated from drinking a diet beverage. This is known as the 'halo' effect, which we'll talk about later in this chapter.

Second, diet soda's higher acidity might throw off the delicate balance of bacteria in your gut which helps regulate both your metabolism, and weight gain. Fowler's main point seems to be that when it comes to trying to trick your brain with fake sugar, there's no fooling Mother Nature. As Fowler told Reuters Health, "Calorie free does not equal consequence free."

Over time, as a natural course for reaching equilibrium and control, your brain's dopamine levels begin to drop. You are no longer impacted as much by your soda intake. This means you find yourself drinking more and more to get the same high you felt before.

The net result? You've just completed a lap in the addiction cycle. If you do enough of these laps, which is to say you've been drinking soda regularly for at least two months, your brain has wired itself for a soda addiction.

According to government statistics, nearly 23 million Americans are addicted to something, and one in ten is addicted to drugs or alcohol. [98] This is because it's not that difficult to hustle our evolutionary reward system. Our culture has developed numerous keys that fit the lock in the brain's nucleus accumbens.

Soda spikes your dopamine levels the same way meth, cocaine, or heroin does, creating an addiction loop. Diet soda works the same way. Tufts University conducted a study of 600,000 people from 1980 to 2010, spanning 51 countries. The study confirmed that 180,000 adult deaths (including 25,000 in the U.S.) could be linked directly to sugar found in a variety of drinks including fruit juice, sugared iced teas, sports drink and soda pop.

What these drinks shared in common was 50 calories or more of sugar per 8-ounce serving. "This is not complicated," Dr. Dariush Mozaffarian, a senior study author told the Los Angeles Times, "There are no health benefits from sugar sweetened beverages," and the net effect could be "tens of thousands of deaths each year."[99]

What do you drink instead? Moderate amounts of decaffeinated coffee seem to help sweep out brain plaques that contribute to Alzheimer's (more on that later). Moderate amounts of wine (one drink a day for a woman, 1-2 for a man based on average body weight) seem to have similar positive effects, due to the polyphenols they contain.[100] Tea has antioxidants that lower the risk of strokes; unsweetened green and oolong teas appear to be especially beneficial. And of course, the safest drink of all is water.[101]

Many of us think of physical health as something totally separate from mental health. However, research suggests it's problematic to think of the brain and the body as separate and distinct entities. A number of chemicals that regulate appetite and digestion in the gut, like insulin, pull double duty in the brain.[102]

In addition to the insulin crossover, neurologists at the Rush University Medical Center report that the protein controlling fat metabolism in the liver can also be found in the brain's hippocampus, where it controls memory and learning. If your fat metabolism gets out of whack, it can have disastrous consequences for the rest of the system. For instance, people who carry a significant amount of belly fat in their forties and fifties are 3.6 times more likely to develop memory loss and dementia later on.[103]

For many, changing an existing diet can be daunting, especially if you've spent a good part of your life creating eating habits that have not optimized your brain functions.

But take heart! It's not all gloom and doom.

The Dark Chocolate Story

Hershey, Pennsylvania may be proud of their namesake, but they have nothing on the ancient Aztecs.

It's speculated that chocolate, derived from the cacao bean (pronounced "ca-COW") had its coming out party some 2000 years ago as an Aztec religious tradition.[104] What you may not know is that some believe the drinking chocolate of the Aztecs began as a beer recipe gone terribly wrong. It seems the Aztecs liked their beer, and enjoyed dabbling in brewing. One can only imagine their surprise, taking that first sip of what they thought would be a rich dark stout, only to discover they'd concocted something closer to an unsweetened hot cocoa.

Intentional or not, a cavalcade of recent medical evidence suggests the Aztecs were onto something. Dark chocolate, defined as 70% or higher in cacao, has some remarkable benefits. Milk chocolate, as in the standard Hershey bar, has much more sugar and much less cacao. But it's less sweet, dark counterpart boasts high levels of both flavonoids and antioxidants. These help maintain health at a cellular level by fighting off free radicals, which have been linked to heart disease and cancer. In addition, cocoa butter is a heart healthy fat with a positive impact on your cholesterol.[105]

Research done in Italy on elderly patients with reduced cognitive skills, as well as a similar study done with patients with good cognitive skills, confirmed that the flavanols found in dark chocolate led to "significant improvements on tests that measured attention, executive function, and memory. ...Both studies found that cocoa flavanols were associated with reduced blood pressure and improved insulin resistance."[106]

While it isn't clear exactly how flavanols work their magic, neuroscientist Miguel Alonso-Alonso suggests they play a role in protecting the brain from toxins and inflammation. Both of which can have compounding effects on overall health.[107]

Sleep

What is the evolutionary purpose of sleep? Doesn't it seem like a huge concession to spend a third of your life inactive? Or is it an amazing strategic adaptation from the days before indoor lighting, when productivity went down once the sun set, and danger came in the form of fast and powerful predatory animals?

UCLA neuroscientist Jerome Siegel suggests the latter. He says that for our ancestors, staying awake and mobile during the evening hours would actually have been evolutionarily maladaptive. We are not the brown bat, who can avoid predators by sleeping 20 hours a day and still manage to hunt blind at night via a sonar-like system called echolocation.[108]

It's been argued that the way our bodies recharge during sleep—not just regaining energy, but rebalancing the immune system and performing brain maintenance—is just another example of evolution piggybacking on an existing adaptation and making the most of it. We aren't equipped for echolocation, so this is what we got instead.

Of course, there is no way to know if Siegel's theory is correct, but it's an interesting idea. What's not in question is that something like 80% of adults regularly get less then eight hours of sleep.[109] And a lack of sleep is unquestionably maladaptive.

Sleep deprivation increases your chances of cancer, obesity, heart disease, Alzheimer's disease, dementia, high blood pressure, stress and depression.[110] If that weren't enough, we now know it compromises your brain's ability to pick up and store new information—in other words, to learn.

When you sleep, the brain sorts through all of the information you've collected in the day, drops out what feels inessential, and retains what it deems important. We know this happens, although we don't understand the precise mechanism of how it achieves this yet.

It appears that your brain assimilates and consolidates semantic memory—facts, figures and bits of specific data—during the first stage of sleep. This means that if you're studying for a test on the periodic table, it's not in your best interest to stay up cramming late into the night. You're actually undermining your retention ability.

During the last stage of sleep, which happens right before you wake up in the morning, your brain builds out episodic memory of physical activities, like learning dance steps or banjo licks, or any other form of performance-based learning. If your goal is to master the latest yoga position, you'll want to make sure you don't wake up earlier than usual and cheat yourself out of a quality downward dog.[111]

A 2011 study done at UC Berkeley suggests that during lower intensity portions of sleep, bursts of brain waves called "sleep spindles" connect to shift fact-based memories from the hippocampus (which has limited storage) to the prefrontal cortex (which has more long-term storage).[112]

The more spindles, the more learning is enabled. These sleep spindles are more active in the second half of sleep, so if you sleep 6 hours or less, it's harder to form long-term memories, which is of course essential for any kind of learning.

Unfortunately, according to Dr. Lawrence Epstein at Harvard Medical School, 20% of Americans get less than six hours of sleep per night.[113]

And it's not just about getting better grades or perfecting your dance moves. The memories you encode and consolidate during sleep form the backbone of your ability to make decisions. Any time we decide something, we weigh our judgments by comparing possible outcomes to memories of previous experiences. Essentially your brain is a giant predicting machine, calculating the odds and laying down a bet.

It's the accumulation of memories that let you, to some extent, predict what might happen next, and this gives you a tremendous advantage in living your daily life.

Hopefully, you only need to burn your hand once on a hot flame to learn and remember the net effect. Memories to a large extent dictate our lives.

Lose your memory to Alzheimer's or dementia and you are sentenced to a harsh existence. If your memories are fogged, you're less prepared to make qualitative decisions, and, when at the wheel of a car for example, that can prove to be deadly.

Sleeping for Safety

Even without a serious condition like dementia, sleep deprivation can make you a danger on the road. If you only sleep six to seven hours a night, you are twice as likely to be in a car accident than someone sleeping eight hours a night. If you sleep six hours a night or less, you are three times more likely to be in a car accident than someone getting adequate sleep. If you work the night shift, you are six times more likely to be in a car crash.[114]

It's a widespread problem; between 50 and 70 million Americans are suffering from one chronic sleep disorder or another.[115] We are a nation of night owls, and a nation of drivers. Sleep-deprived people behind the wheel lead to at least 100,000 car crashes each year, killing an estimated 1550 people.[116] However, since self-reporting is inherently flawed and many people don't even know when they're too tired to function, most experts suspect the real figure is much higher.

The damage isn't just happening on the highway. A 2004 study at

Harvard Medical School found that hospitals could reduce their medical errors by up to 36 percent just by capping doctor work shifts at 16 hours a day, and 80 hours per week.[117] Factor in commuting, showering, and the necessary time to wind down from a shift, and this still doesn't even leave the seven to nine hours of sleep recommended by medical science for peak mental performance. It's just enough rest not to stagger around like a zombie, accidentally adding salt instead of sugar to your coffee—or worse.

NASA has performed extensive studies about productivity—which makes sense, given how much money it takes to put someone into space. Once that astronaut is up there, NASA wants to get their money's worth. NASA's verdict? A 26-minute nap improves cognition by 34-54%.[118] You want to keep your nap at 26 minutes or less, though. Otherwise, you'll screw up your nightly sleep cycle.

On the fence about whether or not you qualify for a nap? Well, do you constantly feel tired during the day, or do you often find yourself falling asleep within five minutes of lying down? Chances are you're not sleeping enough.

The Surrey Studies

Some of the most interesting sleep research today is being done in England, at the University of Surrey's Sleep Research Centre. If you're having trouble sleeping, there are some things you should know.

First of all, your brain powers up as the sun rises. As you encounter sunlight, the brain releases cortisol, a stress chemical that brings you mentally online. As the day progresses, the cortisol is offset by adenosine, a hormone that promotes sleep. (There are several hormones involved in promoting sleep; adenosine is one of the most prominent.)[119] Adenosine continually builds in your system all day long so by the time it gets dark, you feel the need to sleep.

This cortisol/adenosine balancing act worked great, until the invention of artificial light. Electric lighting confounds our brains, which were designed to work with the sun. The trouble is, light bulbs also stimulate your brain's cortisol levels.

That's why one form of torture involves subjecting a person to unending bright light. With a constant stream of cortisol flowing into your bloodstream, it's almost impossible to sleep, which literally begins to drive an individual mad.

Some people are extremely light sensitive. For those people, rolling over in the middle of the night and being exposed to the glowing display of a digital alarm clock can be enough to trigger their cortisol response and wake them up.

The following are some tips for better sleep:

1. Sleep in the darkest room possible.

2. The white light produced by screens—TVs, computers, and smartphones—can also trigger cortisol production. For your best chance at good sleep, avoid every kind of screen two to three hours before you lie down.

3. A cooler sleeping environment can promote better sleep. One way to jumpstart your slumber is to take a hot shower at night. As your body starts to naturally cool down after the shower, the drop in body temperature signals to your brain that you're primed for sleep.

4. The average person needs between seven to nine hours of sleep, so when you're trying to adjust to your own proper cycle, it's a good idea to shoot for at least eight hours. Keep in mind that the eight hours start once you fall asleep, not when you climb into bed. Factor in a little extra time to drift off. If after eight hours, you're still tired, you might need to think about upping it to nine. Each one of us is different so you'll need to do a little experimentation to figure out your sweet spot.[120]

5. Avoid alcohol before you sleep. Initially, a drink or two will make you feel drowsy, but then it will actually trigger brain chemicals that will cause you to wake up.

6. Avoid eating heavy or spicy meals before bedtime. This can lead to GERDs (gastro esophageal reflux disease), in which bile from the digestive system pushes up into the esophagus and wakes you up with a burning sensation in your throat or chest.

7. Also avoid nicotine and caffeine in the evening. They inhibit your adenosine receptors and allow a series of excitatory hormones to party in your head. The net result: you're wide-awake in the middle of the night.

8. As we stated, exercise can lead to better sleep. However, remember not to exercise right before bed; physical activity raises your heart rate and your body temperature, and that's the opposite of what you want.[121]

Brain Maintenance and Brainwashing

Changing the oil in your car or the filter in your furnace is standard operating procedure for maintaining and extending their life.

But what about maintenance for your brain?

You know how, when you exercise, your muscles start to feel like they're burning? That feeling is due to a buildup of lactic acid, the byproduct or leftover waste material from burning glucose in your muscles. Your brain, like your muscles, relies on glucose as fuel. Burning brain glucose also generates a waste product: proteins called beta-amyloids and tau.

We've known for years what happens to the lactic acid in your muscles: your body needs time to flush it out through urination and defecation. That's why after several hours of weeding your garden, your muscles feel sore for a couple of days.

But how does your brain shed the sludge from its glucose burn? For years, scientists knew the waste proteins had to go somewhere, but they weren't sure how it was happening.

In 2013, Dr. Maiken Nedergaard's team at the University of Rochester medical school may have found the answer.[122] According to their research, during sleep your spinal fluid periodically flows up and rinses off the outer surface of your brain, where the unwanted proteins like beta-amyloids and tau collect. (Nedergaard's team studied mice brains, but there's no reason to think the mechanics would be that different in humans.)

This rinsing process can happen at any time of the day, but it appears to be far more concentrated and effective during sleep. During sleep, the flow of spinal column fluid increases dramatically, and the brain contracts slightly, shrinking a little in overall size, allowing more room for the fluid and a more comprehensive rinse cycle.

It's a tidy system, unless, of course, you aren't sleeping enough. Without that rinse cycle—Nedergaard compares it to a dishwasher—those waste proteins can build up on the brain's outer casing, creating what's called "brain plaque," which significantly interferes with neural connectivity. You might think about brain plaque in the same way you think about tooth plaque, where the end result is tooth decay. In much the same way, the presence of brain plaque is believed to be associated with the symptoms of dementia and other types of cognitive decline like Alzheimer's.

There is no workaround. Your brain needs sleep. It is absolutely critical.

Exercise, nutrition, and sleep form the essential foundation for brain health and that all-important willpower. If you don't get those three things in balance, your willpower shrinks, which can leave you forever trapped in *procrastination roundabout* or mired down in *stagnation swamp*.

There is, however, a level of interconnectedness of the brain/body system responsible for willpower enhancement, beyond even the exercise, nutrition and sleep triad we've outlined so far.

Neuroscientists are beginning to understand that the body is governed not by just the brain in our heads, but by multiple auxiliary sub-brains, located in other organs, connected by neurons, and all carefully timed to operate with each other.[123]

The conductor in our scenario would be your internal master clock. Genes in the liver, pancreas, and other tissues each keep time for themselves and coordinate with this master clock to synchronize with the rest of the body,[124] the same way one orchestra section will be both in tune with, and reliant upon, other sections for playing cues.

Caveman at Work

The story of a biologically tuned inner symphony really begins in 1922, when scientists discovered that our little rodent friend, the rat, relies on an internal biological master clock.

Writer Joshua Foer reports that in 1962, 23-year-old Michel Siffre literally brought the master clock story to the light of day.[125] To prove humans, like rodents, also had a biological master clock, Siffre set up an experiment. But not just any experiment, this one involved him proving his point by descending into a cave and living in complete isolation for 63 days.

Into the cave Siffre went, along with food and supplies. His cave had a telephone, allowing him to call his monitoring team above ground and report when he awoke, ate, and slept, as well as his pulse rate during each of these events.

He had no way to tell what time it was, since he had no watch or clock. His world was lit by a single light bulb, and, as part of the experiment, his support team was not allowed to call him with any news from the outside world, including what time or day it was.

Absent the cues of any time-keeping mechanism, including daylight or evening's darkness, his body still fell into a natural cycle of about 24.5 hours. He claimed he experienced 'perfect sleep.' (It should be noted that in subsequent cave experiments, occasionally the participants would fall into a 48-hour cycle: awake for 36 hours followed by 12 hours of sleep. Why this occurs is still not understood.)

One of the experiments Siffre conducted involved a counting test where he attempted to count from 1-120 at the rate of one digit a second. The longer he was underground, the more distorted his counting became.

Eventually it took him five minutes to reach 120.

Another distortion of time occurred when the experiment concluded. Siffre entered the cave on July 16 with the intention of finishing the experiment on September 16. When he was told the experiment had concluded on September 16, he thought it was only August 16. Foer reports, "his psychological sense of time had been compressed by a factor of two."[126]

So, although our psychological sense of time can be affected by taking away external cues like light and darkness, the cave experiments proved our internal biological clock hums along quite nicely unimpeded by the outside world.

Without knowing it, Siffre's early experiment made him the unofficial father of chronobiology, the study of the brain's internal biological timing mechanism. Siffre's experiment helped prove that, like lower mammals, we too have a master clock governing our living processes.

Your Master Clock

Today, we know a master clock resides in the brain's suprachiasmatic nucleus,[127] or the SCN, which helps us govern physical, mental and behavioral changes, all following a fairly consistent 24-hour cycle.

The SCN contains as many as 20,000 nerve cells and is located in the hypothalamus, just above part of your eye circuitry known as the optic nerve. This location is pretty handy because the SCN relies on signals from the eyes regarding daylight and the dark of night. This, in turn, helps regulate your circadian rhythm and the release of hormones like cortisol to wake you up and adenosine to make you drowsy for sleep.

According to a Sleep Foundation report, our circadian rhythm tends to fluctuate throughout the day.[128] In the middle of the day, we often feel sleepy, but many of us push through that feeling, helped by a caffeine hit from coffee, tea or soda pop.

How does sleep fit into your circadian rhythm?

Most sleep experts agree that seven hours is the bare minimum you need before you begin to incur a sleep deficit. Sleep deficit is the cumulative lowering of your functioning over time. Stimulants like caffeine, brightly lit work areas and physical activity can mask the degradation in performance that comes from perpetual sleep loss. Thus, people often don't recognize the debt they incur by repeatedly cheating their sleep cycle.

A "typical" circadian rhythm is probably an oxymoron; we all run by a slightly different clock. The precise timing depends on when you wake up and go to sleep, but certain events happen in predictable intervals. For a rough idea of what that clock looks like, we might imagine an adult who wakes up at 6 a.m. and goes to bed around 10 p.m.[129]

It might look something like this. At 6 a.m. light streams through your window, prompting your brain to release a shot of the hormone cortisol to help wake you up. Not catching enough sunlight? You might supplement that cortisol with a little friendly caffeine, in the form of a steaming cup of joe. That grumbling in your stomach is a signal that your metabolism is back on-line and ready for more fuel after its long evening fast.

By 7 a.m. adenosine, the hormone designed to sedate you for nighttime rest, is no longer present in your bloodstream. The message is clear: there's a brand new day waiting for you.

Around 9 a.m. your sex hormones peak. (Of course, what you do with that information is up to you.)

By 10 a.m. you are at your daily high point in terms of mental alertness. If there is some critical thinking challenge ahead of you—an important deal to cut, a spreadsheet to build, a whitepaper to write, a speech to make—carpe diem. Your metabolism is starting to gear up and right before noon, it's at its peak.

At noon, you take in lunch, and your digestive system gets busy converting the food you've consumed into useable fuel in the form of glucose. This affects your metabolic rate and your blood sugar levels. Just how it affects you somewhat depends on the contents of your plate.

If you've eaten a lot of carbs, by 1:30 p.m. you may be feeling a little tired and cranky.

After your digestive tract gets a chance to work its magic, you push through that post-lunch dead zone, and by 2:30 p.m. you're starting in on your second wind. This coincides with the peak of your motor coordination. If you're thinking about any intricate physical movement, from practicing the accordion to hitting the ballroom to polish your paso doble, the moment has arrived.

When 3:30 p.m. hits, you also hit your stride as far as reaction times go. Taking a little batting practice? Trying to get to the next level of Legend of Zelda? It won't get much better than this.

By 5 p.m. your cardiovascular system and muscle strength are at their zenith. Now's the time to go for that bench press record, or challenge your officemates to a race around the parking lot.

When 7 p.m. rolls around, your blood pressure and body temperature are at their highest, much like a car engine that's spent the better part of the day burning up miles on a cross-country road trip. Your brain, however, is not at tip-top: fatigue has set in, and the quality of your decision-making has eroded by 70% compared to that 10 a.m. sweet spot. The level of adenosine in your blood is beginning to gain steam.

At 9 p.m. your body starts preparing for sleep, provided you've backed off your screen time. Your metabolism and digestive system slow and your bowels begin to rest, making sleep more comfortable—and, given the alternative, more hygienic.

When darkness falls, your brain revs up its adenosine production, so that by 10 p.m. there is enough of this chemical in your bloodstream to subdue physical movement and conscious thinking, thereby ushering you into a night of peaceful sleep. Your metabolism is on autopilot, and eating at this point will likely default those calories into fat storage. Adenosine is busy pumping away in your blood. The message is less than subtle; you feel the need to sleep.

At 2 a.m. you are in the deepest part of your REM sleep cycle. If you wake up during this time, the result is grogginess, confusion, and a huge toll on your brain's memory consolidation process.

When 4 a.m. arrives, your body has fully powered down. You experience mini- hibernation. You've cooled down to the point where you might find yourself pulling up the covers to squeeze out those last few hours of sleep. The stress hormone cortisol begins building in your system, anticipating the need to wake up.

Two hours later, sunlight streams through your window and that new shot of cortisol wakes you up. Another 24-hour circadian revolution draws to a close, a new one begins, and the cycle continues.[130]

A couple of things to keep in mind: if you're an early riser, your circadian cycle will start up earlier, which moves up the timing on everything. If you're a late sleeper, your pattern will be extended, moving everything back a little later in the day.

Neuroscientist John Medina classifies people into three basic sleep types: early-rising "larks" who are most productive before noon, late-rising "owls" who are most alert around 6 p.m. and the rest of us: "hummingbirds" who follow the middle-of-the-road pattern typified above.[131]

Your Circadian Cycle

There is no one-size-fits-all solution, but understanding your own circadian rhythm can help you improve your personal performance. There are a couple of keys to remember in order to take advantage of your internal clock.

Depending on what skill you're trying to conquer, you can see it might make sense to time-sequence your activity. For instance, it's not your best move to tackle a difficult writing assignment late in the evening.

It's also crucial to try to maintain a constant sleep schedule, particularly getting up at the same time every morning.

Traveling by plane and crossing time zones can throw off your biological clock. The effects of jet lag can take its toll on your alertness, the quality of your decision-making, your reflexes, your mood, and your memory. It's estimated that you need one additional day to recover for every hour of time zone change you encounter.[132]

If you're not a world traveler, you could still be experiencing similar effects if you do shift work—that is, jobs that operate outside of the standard nine to five schedule. According to the National Sleep Foundation, between 10 to 20% of shift workers report having fallen asleep on the job.[133] And sleep deprivation in the workplace is no laughing matter. "Historic tragedies have been linked to fatigue-related human effort," the Foundation notes, "among them, the Exxon Valdez oil spill, and the NASA Challenger shuttle explosion."[134]

If you're tempted to make up for lost sleep over the weekend, then you should know that unfortunately, sleeping in on the weekends has the net effect of throwing your circadian rhythm further out of sync. Missing out on your proper sleep cycle, and thus ignoring the circadian clock's cues of light and dark, is akin to the orchestra conductor not showing up for work.

An erratic sleep schedule doesn't just leave you dozing on your coffee break. Critical brain neurons are designed to operate in the pre-dawn to prepare your heart for the rigors of the day. Throwing off their operations is bad news for your heart. It's theorized that this is why so many heart attacks occur in the early morning.

Additionally, when the adipose layers of the body—the fat storage system—get off their natural circadian rhythm, they can release fatty molecules at the wrong time of the day.[135] This plays havoc with your metabolism and causes you to gain weight. Your pancreas relies on the same clock system to release insulin; if that routine is repeatedly disrupted, the result can increase your chance for diabetes.

If you've ever heard an orchestra tuning up before a concert, you begin to appreciate just how important the conductor really is to keep everyone on the same page and the same beat. Even the greatest musical ensembles have nothing on the human body. Staying in tune starts with the sun, runs for roughly twenty four hours, and relies on a whole interconnected collection of biological timekeepers oriented to a master clock, ticking forward every moment without the benefit of gears or batteries.

What About Caffeine?

Americans love their coffee. According to the National Coffee Association, we drink an average of 3.1 cups a day and spend a total of $40 billion every year on it.[136] We depend on our caffeine, for a boost in the morning or for an after-lunch second wind.

The buzz you feel after slamming a latte is not actually the caffeine.

Remember adenosine? It's the circadian rhythm hormone that helps usher in sleep, starting as a slow trickle when you wake up in the morning and slowly building in your system so that by nightfall, you're ready for a little shut-eye. Caffeine interferes with the adenosine receptors when it hits your brain.

When the adenosine supply is cut off, the brain's naturally occurring stimulants, dopamine and glutamate, act like a couple of kids with mom and dad on vacation, which is to say: they party. That surge of energy after a cup of coffee is just your dopamine and glutamate going wild.

Caffeine hits your adenosine receptors within fifteen minutes of ingestion and remains in the bloodstream for a full 24 hours. It has what's called a six-hour half-life, which means a cup of coffee in late afternoon will still retain half its potency by bedtime. Although it might give you a temporary pick-me-up, enough of it can greatly disrupt your REM cycle sleep, leaving you emotionally spent the next morning. The typical solution? Another cup of coffee.

Once your brain catches onto your caffeine pattern, it starts producing more and more adenosine receptors to offset the caffeine. This means you have to keep ratcheting up your coffee intake to get the same high. You are now in an escalating mini-arms race of chemicals battling within your own brain.

Caffeine is like any other addictive drug. In "Caffeine: The Silent Killer of Emotional Intelligence",[137] Travis Bradberry explains what happens when people try to quit: "The researchers at Johns Hopkins found that caffeine withdrawal causes headache, fatigue, sleepiness, and difficulty concentrating. Some people report feeling flu-like symptoms, depression, and anxiety after reducing intake by as little as one cup a day."

Do you use caffeine as a wake-up call in the morning? Interestingly, you can get exactly the same effect by exposure to bright sunlight, thereby inducing additional cortisol to bring your brain online. Provided you have access to sunlight, you can save yourself the downside of caffeine's half-life effect and maybe a couple of bucks at Starbucks every morning.

But hold on a second, says Gary L. Wenk in *Your Brain on Food*.[138] Coffee itself is an excellent source of phenols, substances that function as both antioxidants and anti-carcinogens. He cites preliminary studies that suggest drinking 3 to 5 cups a day (decaf counts) has a positive effect on Parkinson's disease, prostrate disease, diabetes, and certain cancers. He notes that two or three espressos can also have a similar effect.

Adding Sleep

Should we try to sleep more than eight hours if possible? What are the net results of increasing sleep beyond the standard recommendation of seven to nine hours? Should, say, elite athletes try to maximize their performance by catching some extra Z's?

To answer these questions, scientists performed an experiment on the Stanford men's basketball team. Prior to the study, they measured the team's skills in three demonstrable areas: free throws, three-point shooting, and finally, each player's speed in an 80-meter dash.

Then for five weeks, the players were allowed to sleep as much as they wanted. Over the course of the experiment, the median sleep time turned out to be a little over 10 hours a night. The results? After five weeks of extended sleep, on average, the players improved their free throw shooting by 9 percent, their three point shooting by 9.2 percent, and their individual 80-meter sprint times by .06 seconds.[139]

Even though the test was a rather small sampling group, it does raise important questions about how much sleep is required for those individuals undertaking tasks with unusual physical or mental demands. Another unanswered question: what does additional sleep do to your circadian cycle?

We are finding that the human system is far more complicated than we ever imagined. In the end, much like love, real estate, and yes, music, it's all about timing. And there is no escaping a simple truth: "sleep is not a luxury, it's a biological imperative".[140] You just can't reach your highest levels of skill or performance without it.[141]

Boosting Willpower

We now know that the triad of exercise, nutrition and sleep are the foundational cornerstones of willpower. Is there more to it than that? In Chapter One, we learned that early talent is no predictor of success— and we've talked about why it isn't— but is there any data indicating which of us are more likely to succeed toward our long-term goals? The simple answer: yes.

Although he didn't realize it then, when physiologist Walter Mischel and his research team turned their attention to preschoolers at the Bing Nursery School on the campus of Stanford University in the 1960's, they were about to unleash a tidal wave in cognitive science that's still being felt today.[142]

The experiment was straightforward. A preschooler was seated at a table and presented with a single treat of some kind—a cookie, a piece of candy, or—most famously—a marshmallow. The child was told that the supervisor would soon be leaving the room. If the child could resist eating the marshmallow until the supervisor came back, he or she would get a second marshmallow. If the child instead ate the marshmallow, there would be no opportunity for a second treat.

Mischel was studying delayed gratification, and whether five-year-olds could demonstrate any kind of strategy for self-restraint, the willpower necessary not to gobble up that first tempting marshmallow.

As you might imagine, some children ate the marshmallow almost as soon as the supervisor left the room. Some ate the marshmallow before the supervisor even got out the door. But some were able to hold out for the future reward of another marshmallow, even for as long as 15 minutes. To do this, the children employed several different methods. It turns out that by four years old, some children have already developed fairly sophisticated delayed gratification techniques.

Among the participants were Mischel's own three young daughters, and a number of their classmates. Over subsequent years, dinner conversation at the Mischel table included his daughters catching him up on how little Sally or Billy was doing in the years since the test.

As Mischel's daughters shared their observations, Mischel started to detect what seemed like a pattern. According to them, the kids who had resisted the lure of the first marshmallow appeared to be faring better in school and bonding more with their classmates than the gobblers.

On a hunch, Mischel launched a longitudinal study. Ten years after the experiment, his team tracked down the original participants and measured their success and happiness across a number of areas. They were measured every ten years afterwards for the next thirty years.

In his book *The Marshmallow Test*, Mischel explains how, a dozen years after the test, those who had "exhibited more self-control yielded less to temptation, were less distractible, were more intelligent, self-reliant, and confident and their SAT scores were significantly higher. By age twenty-five to thirty, those who had delayed longer in preschool self-reported less risky drug use, higher educational levels, greater income, had a significantly lower body mass index, and divorced less."

This experiment has been replicated many times all over the world—and not just because there is something innately hilarious about watching tiny children struggling to hold themselves together in the face of a single sweet. Again and again, researchers have come to the same conclusion: those who adapted and adopted early strategies for self-restraint fared better than their counterparts in a variety of ways over the course of their lives.

Not sure you would've been a two-marshmallow kind of kid? Don't worry, says Mischel. Willpower, or more accurately, a strategy for maintaining one's willpower, is a skill that can be developed and practiced at virtually any age.

The clues to building your own willpower strategy can be found in Stanford's Bing Nursery preschoolers, who tested and proved many methods during the original marshmallow experiment. The next time you need a little more self-control, you might consider using one of the following as a template.

Out of Sight, Out of Mind

Among the marshmallow resistors, pushing the marshmallow to the far end of the table and/or closing their eyes was a popular way to ramp down the emotional brain's impulses. Our brains register all information through our senses; fewer sensory inputs mean a reduction in cravings. Out of sight, out of mind.

If you're trying to eat healthier, stop the waiter before he even brings you the basket of freshly deep-fried tortilla chips or wheels out the dessert cart. Make things easy on yourself.

Reframing

Children in the experiment had a much better shot of holding out if the researcher asked them to use their hands to frame the marshmallow like a picture, rendering it unreal in their minds.

One child was quoted as saying, "Well, you can't eat a picture." As silly as this may sound, he had a point. Cravings exist, not in objective reality, but inside the confines of the emotional brain. Reframing can actually lower the potency of the chemical surge in the amygdala that urges one toward grabbing that tempting marshmallow.

Association

Here is a little more mental jujitsu at work.

Your reflexive System 1 process is designed to make snap judgments by storing a huge number of unconscious links between ideas or images based on past experience. You can leverage this natural predisposition by imagining that a cockroach has, only moments before, been living inside the marshmallow, or that piece of chocolate cake was dropped on the restaurant floor before it landed on your plate. Creating this false association can put a real damper on emotional desire.

The same principle explains why many college freshmen, after getting drunk and vomiting up the food they just ate, will swear off that food afterward. You may know someone who will readily testify how these unwanted associations can leave an indelible mark on food decisions for a very long time.

Distraction

What doesn't appear to work is to simply tell yourself that you're *not* going to think about the marshmallow. It's like that old joke, "Whatever you do, *don't* think about the elephant in the room." Fixating on something puts the image more firmly in your working memory and can actually compel you toward indulgence. There is a name for this phenomenon; it's called, "ironic rebound." Psychologist James Erskine of London's St George University has done experiments with chocolate and has proven the more you try to resist thinking about chocolate, the more chocolate you'll eat, and as his experiment demonstrated, frequently as much as double your normal intake.[143] The key to abstinence is to get your System 2 analytical process focused on something else entirely.

In the end, the one important thing that the two-marshmallow kids demonstrated was strategy. Simple or goofy as their solutions might appear, the skill of delayed gratification led to real-world benefits that lasted the rest of their lives.

Without a strategy, a lack of willpower can become a major stumbling block, thwarting our long-term goals, dooming us to *procrastination roundabout* and passing the buck to Future-Self.

Willpower and Focus

Ask anyone you know, "What long-term goal do you want to work on, but haven't managed to tackle so far?" and no matter who it is, they'll have an idea. And if you ask them why they haven't gotten rolling yet, most of them have an excuse, or they'll invoke Future-Self. Some, undoubtedly, are unaware of the importance of balancing their sleep, nutrition and exercise to build maximum willpower.

And some will say, "I just couldn't seem to stay focused. I worked on it for awhile but I couldn't keep it up."

In her book *The Willpower Instinct,* Stanford's Kelly McGonigal, equates willpower with three separate systems within your prefrontal cortex. You might recall that the prefrontal cortex serves as headquarters to analytical System 2. McGonigal says these subsystems are designed to overcome your more primitive System 1 impulses. She jokingly named these subsystems the "I will, I won't and I want system." (Anyone that's ever raised a two-year-old is familiar with these concepts.)

The "I will" system, located in your left prefrontal cortex, pushes you forward on your goal. When you lace up your running shoes for a jog even though you're tired from working all day, that's your "I will" system in action.

The "I won't" system in your right prefrontal cortex resists temptation. That's the resilience the two-marshmallow kids repeatedly demonstrated in Walter Mischel's famed experiment.

The "I want" system found in your ventromedial prefrontal cortex forms a kind of bridge between your executive control center and your more primitive, reactive processes. It's designed to regularly remind you of the big picture, and to help keep you focused on your long-term goal. (In other words, it whispers that there are two marshmallows available if you can only hold out.)

McGonigal suggests that every time you demonstrate willpower, you are engaging one of those three subsystems. And the good news is, the more you engage any of them, the more you reinforce that neural wiring, and thus, the more powerful your subsystems become.

McGonigal's research supports what we learned in Chapter Two about strategy: willpower depends on setting clear goals, attacking those goals incrementally and understanding the learning curve attached to each goal. She points out that one of the easiest ways to hack that system is through meditation. More about that in Chapter Seven.

Beware Willpower's Moral License

So far we've described the upside of willpower, but is there a downside? Neuroscientists coined the term 'halo effect' to describe the irrational behavior that follows when we grant one positive element too much sway in our decision-making.

According to McGonigal, the halo effect is a "form of moral licensing which allows us to say 'yes' to temptation."[145] Exercising our willpower in one case can cause us to get complacent and slack off in another. "Researchers have found if you pair a cheeseburger with a green salad, diners estimate that the meal has fewer calories than the same cheeseburger served by itself," says McGonigal.

It's as if there's a bad angel sitting on our shoulder, waiting to balance the virtuous salad we had for lunch against a Quarter Pounder with cheese. The heavenly light reflecting from our mixed greens casts a glow over the rest of our plate, which magically makes some of the calories disappear.

Based on this illusion, diet soda is like a gateway drug to high fat and high sugar indulgences. After all, you saved all those calories on your beverage; surely you're entitled to spend some of your saved caloric capital elsewhere. As McGonigal explains, "The Snackwell cookie craze of 1992 is a perfect example of this kind of moral licensing. When dieters saw the words 'Fat Free' on the outside of the package, people watching their weight irrationally consumed whole boxfuls of the high-sugar treats, blinded by the light of the fat-free halo."[146]

Again, we can see that the brain's decision-making is not always rooted in rational thought. Marketers long ago recognized this and have for years exploited us with manipulative words like 'fat free,' and 'sugar free.'

This cheap ploy even works with people who are well aware of the calorie count for many foods. Logic be damned, they succumb. What seems like a good choice can end up being the slippery slope for your future food gorge-fest.

The Power of Glucose

If you run up a long steep incline, it doesn't take very long before you burn through the energy stored in your muscles and find your legs turning to rubber. We learn this at a relatively early age, and as a result, some make it a habit to avoid running up steep inclines.

This exhaustion, this depletion of fuel, happens similarly when you exert yourself mentally. As we've learned, your brain, like your muscles, runs on glucose. Give your brain a mental workout, and your ability to focus, or demonstrate what we call willpower, is spent as well.

This was proven in a well-known experiment done by psychology professor Roy Baumeister and his team at Florida State University. They conducted a test in which people were randomly assigned to eat either radishes or freshly baked chocolate chip cookies. The radish eaters were instructed to resist eating the cookies. In this case, the noble radish eaters were able to exert enough willpower to avoid the cookies 100% of the time.

Both groups were then presented with a series of problems that required extreme eye and hand coordination. The radish eaters gave up noticeably sooner than the cookie eating crew — the radish eaters' focus was about 10% shorter.[147]

In this and other experiments, Baumeister was able to show that the more often and more recently you resist a desire, the less likely you're able to complete the next tough task that comes along. This might explain why you were able to bypass those cookies your workmate brought in at lunchtime, only to find yourself with no energy left to scrub the bathroom floor that evening when you got home from work.

Displays of physical and mental energy deplete your willpower in the same fashion; you only possess one store of glucose, shared by both your brain and body.

Luckily, people can choose to conserve their glucose and hang on to some willpower, and so for this reason, we are not necessarily reduced to mush after a series of extremely tempting situations.

When it comes to willpower, the muscle analogy holds up: the more you exercise it, the better you grow at resisting whatever temptation might befall you. And all of this is necessary to stay in pursuit of your long-term goal.

Remember Dave Guy, our trumpet player? We know that Dave was on a mission to become the best musician he could be, and his growth mindset pushed him to risk short-term failure in order to learn and improve. What kind of willpower did Dave display?

Sometimes the low points in our lives, while at the time painful and difficult, can usher in opportunities that we might not otherwise recognize. Early in Dave Guy's music career, his girlfriend broke up with him. Dave did what many of us do in that kind of situation: he paused to take stock. He decided it was time to get himself in shape. He started working out.

About this same time, Dave became more concerned with what he ate, and started to take in more proteins and vegetables and reduce his intake of junk carbs. Without knowing it, Dave was rebuilding his exercise-sleep-nutrition triad. He was boosting his willpower up a notch and incrementally moving toward his long-term goal.

Chapter Key Points: Hacking your Willpower App

- Exercise, nutrition and sleep, are not three separate patterns. They are one system supporting what neuroscientists call the mother of all virtues: willpower.

- Exercise allows brain cell walls to stockpile additional glucose, improving your decision quality, stamina and blood flow.[148]

- For the highest level of energy (both mental and physical), build your daily food intake around greens, healthy fats, and lean protein. Keep an eye on your caffeine and sugar intake.

- An eight-hour sleep cycle is crucial for a balanced immune system and memory consolidation. It also increases focus and tenacity, which allow for greater willpower.

- For an added push toward your goal, understand strategies for delayed gratification, like 'out of sight, out of mind', reframing and association. Just beware of moral licensing and the 'Halo effect.'

- Demonstrating willpower will positively spill over into other areas of your life.

Chapter 4

Deliberate Practice App

Let's assume at this point you're operating with a growth mindset. You've planned for your long-term goal and you're taking advantage of exercise, nutrition and sleep to boost your will power and keep you on track, steering clear of *procrastination roundabout*. Future-Self has left the building.

What's our next move?

The Pit of Practice

No matter what your long-term goal, building skill is a necessity for success. How do we hack into the brain's practice schedule? Let's take a quick look at how we develop skills.

Brain structure is often described as an immense tangle of electrical/ chemical connectors called neurons. These neurons have the ability to connect and disconnect, allowing us to drive a car, gargle after brushing, watch TV, read Shakespeare, play chess, and perform all the other tasks we do, including playing the piano, if indeed you play.

Even if you don't play, just thinking about playing the piano is making your neurons work. Neuroscientist Daniel Levitin tells the story of his friend Petr Janata, using super fine wires to hook up an owl's brain neurons to a loudspeaker. When he played music, the owl's brain produced a series of audible clicks as its neurons fired in time with the music.[149]

Levitin says this is possible because the output of a brain neuron is about 70 millivolts—comparable to the current of an iPod. This neuron clicking activity is frequently compared to a series of switches flipping on and off. In humans, neurons do this in mass, sometimes firing in clusters of 10,000 or more at a time.

This image—a myriad of switches, waiting obediently and drone-like for activation—is something we can envision. Although the sheer numbers and speed of neural operation is difficult to contemplate, this concept certainly seems easier to picture than, say, the theory of relativity or quantum mechanics.

Learning something means you're rearranging the neuron connections, and building new neural formations. Neurons are like a giant box of Legos, which can be repurposed over and over to create extraordinary new mental combinations. So if you pick up any new information from this book, you've literally altered the structure and wiring of your brain.

Imagine if each time you used your GPS, it worked a little faster, and with improved accuracy. From this angle, even the most cutting edge technology lags far behind the ancient electrical/chemical signaling system we all carry in our heads. We program and re-program ourselves every day.

This means that the more trumpeter Dave Guy activates the neural circuits that govern trumpet playing, the more neurons learn to activate together and bond, thereby both altering and increasing the signal bandwidth and speed. This is known as Hebb's rule, sometimes paraphrased as "neurons that fire together, wire together."[150]

Our brains are the sum total of what came pre-wired at birth plus connections we've created through learning. Researchers can actually track the accumulation of skill by using an fMRI to compare structural differences in a brain before and after a practice regimen.

When you first engage a new neural pathway, your brain is firing at a rate of two miles an hour. You can walk faster than that initial neural signal. Of course, the message only has to travel a short distance. In most of us, from tip to toe, it's roughly between five and six feet, but the fact remains: two miles an hour.

When you reuse the same circuit enough times, and take the time to build flawless code, the power of those electrical/chemical signals increase, ultimately moving at a rate of 268 miles an hour or more.[151] At this point in his life, when Dave Guy is improvising jazz on his trumpet, his circuits are blazing away at over 268 miles an hour. A dirt road has turned into a super highway.

Building Bandwidth: Myelination

The more times you perform a given action, the more your brain insulates that neural connection with a fatty substance called myelin. Myelin protects and reinforces neural signal strength through containment, just as electrical wiring in your house is coated with protective insulation.

(Diseases like multiple sclerosis involve the sloughing off of the brain's myelin insulation, which causes electrical signals to weaken and dissipate. Unfortunately, in the case of multiple sclerosis, de-myelination shows up as a loss of motor skills.[152])

That's why repetition is so vital to learning: it fosters new layers of myelin insulation. When you are learning a new skill and repeating an action over and over again, you're myelinating.

Our ability to myelinate is at its peak when we're young. In most of us, by the time we turn fifty, myelination is on the wane, but if we keep up the willpower triad of nutrition, exercise and sleep, we can continue to myelinate as long as we live.[153]

This ability to build and strengthen neuron connections allows you to develop and run mental programs that make your life easier: what we call brain apps. It's estimated that just under 50% percent of what you do every day is reliant on these routines, pre-programmed through repetition. For example, turning on a light switch in a dark room doesn't require conscious thought. You've built that neural program long ago and your unconscious brain runs it automatically, no conscious thought involved. This is why, if your home loses power, you *still* find yourself automatically reaching for the light switch.

These shortcuts are invaluable to humans, and modern technology would have little meaning to us without them. Unconscious memory is like accessing the owner's manual for everything you've learned to use over the course of your lifetime, from appliances to automobiles.

Some apps are preprogrammed at birth—breathing, for instance—while others you had to intentionally create. Take driving a car, for instance: at one point when you first learned to drive, you had to actively remind yourself to apply the gas and the brake pedal. But after going through a Driver's Ed program, logging the practice hours, passing a test, and continuing to drive your car on a regular basis, it's unlikely you'll wake up tomorrow morning and say, "Oh my gosh, I've totally forgotten how to drive."

These days, you don't even need to think about starting the car—your hand floats up to put the key in the ignition or push a button when you climb in. You can get into your car, drive twenty miles, and remember almost nothing about the individual steps you performed behind the wheel; your awareness wasn't needed. Your driving app was running in the background, the same way programs run in the background on your computer.

Especially if you live in a place without good public transportation, the driving app can be invaluable, but other apps are insidious. We've all programmed ourselves with at least a few bad apps: smoking, overeating, and—one of the most insidious—over-reliance on Future-Self.

Your brain doesn't sort out the good apps from the bad ones, and here's why: the basal ganglia, that wad of tissue in the middle of your brain that helps encode behaviors, doesn't pause to weigh the ramifications or apply a value judgment. It will wire you to water the plants every day as easily as it will wire you to bite your nails. The brain is aiming for maximum efficiency, and so anything identified as a frequent action gets reinforced, and will eventually be operated unconsciously.

Most of us are not accustomed to thinking that we can consciously decide which brain programs we're going to create and run. But in many ways, it's entirely up to us. It generally takes about two months of repeating a simple activity before a brain app gets built to the point of automation. More complicated tasks like perfecting your driving skills can take longer.[154]

Repetition: A Key to Skill Acquisition

On a technical level, skill is just an accumulation of myelin around particular neural circuits, driven by those all-important practice reps. For instance, what Tiger Woods did—or rather, what his father and coach Earl Woods did, since two-year-old Tiger probably wasn't aware—was to systematically wire Tiger's brain for the best possible golf performance. It's a way of learning unfamiliar to most of us. How does it work? Through something *Talent Code* author Daniel Coyle calls "deep practice,"[155] more commonly referred to in the scientific community as *deliberate practice*.

Deliberate practice is different than play or sheer repetitive learning. When we engage in play, like a relaxed round of golf or a pick up basketball game with our friends, we are reusing the connections we've already forged. It's similar to driving down a familiar road. We aren't improving the road, we're just wearing the established path a little deeper. The mere act of accessing that golf or basketball app and hitting the play button doesn't actually improve our skills. We are simply preserving those established myelinated pathways in our brain.

By the same token, if we actively practice the same routine over and over again, whether it be hitting a golf ball or shooting a jump shot from the top of the circle, or even going through a sales presentation, there is a point when the repetitions will cease to yield additional benefit. Our improvement in these scenarios will be logarithmic, and will start to taper off and eventually stagnate. Rote repetition is not the same thing as deliberate practice.

Both playing for fun, and repetitively performing a previously myelinated activity might bring us some level of enjoyment. Your best friend might get into playing the opening bars of Hungarian Rhapsody over and over again on her electric violin, but if she's been doing it long enough, and let's face it, you've heard it plenty of times to know this, it doesn't sound any better than it did when she was playing it four years ago. It's easy to get caught in this kind of stagnation trap where we treat repetition and practice as one in the same.

There are two important maxims to keep in mind about deliberate practice.

First, if you're having fun doing it, or you find yourself defaulting to retreading the same actions, you are probably not engaged in deliberate practice. You might enjoy what you're doing, but you're not actually improving. This is exactly where many of us find ourselves. We get to point where we have some level of competence, like the ability to play part of Hungarian Rhapsody, but that's essentially as far as we ever go. The Yo-Yo Ma's and Mark O'Connor's of the world didn't stop there; they continued to hone their playing skills.

If your goal is to improve far beyond what's typical, then you have to be much more deliberate and creative about the way you practice. Deliberate practice involves an extraordinary amount of System 2 analytics and dedicated concentration. It requires you to operate right up to the painful edge of your mental focus.

In fact, maintaining this focus burns your glucose levels at such a high rate that you tire pretty quickly. One of the telltale signs that you're practicing deliberately is feeling spent both mentally and physically early in a practice session. When engaged in deliberate practice, depending on which skill you're trying to acquire, you might find yourself running out of gas after as little as 10 minutes. In most cases, three hours is the very upper limit of how long you can sustain deliberate practice[156]. After that, your mind and body need downtime to recuperate.

Deliberate practice, as we will see, is a very specific way to acquire a skill. Some may show early promise on the field of talent, but deliberate practice is generally a prerequisite for the extraordinary talent we label 'genius'. Granted, it doesn't sound nearly as exciting as the heavenly glow of spontaneous brilliance we attribute to the Tesla's and Einstein's of the world.

Confirmation Bias

Just how good are you at music, selling, sports, art, math, or acting? How much improvement are you making day in and day out? Self-assessment is difficult for one very simple reason: humans have a built-in confirmation bias. We tend to overestimate our skills.

Some scientists believe that biasing toward a higher level of self-assessment allows you to maintain a positive attitude, and thus can help you overcome daily setbacks.[157] But as often is the case, too much of a good thing isn't always good for you.

In one study, college professors rated themselves on the quality of their lectures. Ninety-four percent rated themselves above average.[158] Clearly, that's numerically impossible; everybody can't be above average. This confirmation bias goes hand and hand with the Illusion of Depth[159].

Illusion of Depth

If confirmation bias is the belief you're better at a task than you really are, Illusion of Depth lets you think you have more knowledge than you really do. You can see how these two biases can create a powerful double whammy and keep you treading water instead of reaching your goal.

The antidote to these biases is an investment in domain knowledge. Domain knowledge is a deep understanding in a given field. Famous twentieth century inventor Nicola Tesla's comprehensive knowledge of electromagnetism led to the creation of the alternating current electrical system still in use today. Every time you plug in your smartphone or hair dryer, you can thank Mr. Tesla.

Imagine that you want to be an expert, and, like Tesla, build a deep well of domain knowledge. To understand the process, we can break down the pursuit of domain knowledge into four phases: learning, practicing, mastery and implementation.

First we begin with learning, the introduction to new information in a given domain. Maybe you're attending a seminar on sales skills, and in the learning phase, you've been introduced to a brand new sales closing technique.

If, after attending the seminar, you feel like you have a really clear grasp of the closing concept, your brain's next move is to engage in a subtle deception. You begin to imagine yourself using the new closing technique—and not just using it, but likely excelling at it.

As we've learned, a positive imagined state is foundational for growth mindset, but it can lead to the illusion of depth, mistaking the imagined state for a level of mastery.

In this imagined state, you might be tempted to bypass phase two, the practice phase, wherein the skill is actually acquired. It's important to remember that learning about something is not the same as mastering it. As simple as this idea sounds, it's often neglected, and not just by individuals. Many companies and corporations, routinely devalue the role of practice toward skill mastery.

Watching a PowerPoint presentation or a video on YouTube or Vimeo might expose you to new information, but doesn't guarantee true learning, transfer and mastery. This is the port from which many a ghost ship has sailed. There is just no shortcut to reinforcing layers of myelin through practice.

Eventually we come to phases three and four. Here, you'll need some kind of assessment to prove that you have, indeed, mastered and can perform the skill. As much as possible, skill testing should be in the context of real life conditions. Playing a piano recital in front of a live audience is very different than practicing specific fingering and painstakingly working through the musical scales as a metronome paces you. And this, in turn, is very different from playing the piano by yourself in the comfort of your living room while your cat looks on disinterestedly. Performing a successful mistake-free recital in front of an audience is one way to demonstrate the necessary skills have been mastered.

Remember, myelination is the gateway to mastery, and without proper practice, the result can be fear-driven mistakes, or overthinking. Some of us have painful childhood memories involving sporting events or music recitals, where in the middle of a less-than-rehearsed performance, the brain mutinied against the hands or feet. This is often called 'choking'.

Choking

During the 2012 Olympic games, the U.S. men's gymnastics team gave an unintentional clinic on choking.[160] Despite its lack of a head, or any moving parts, the men found the pommel horse nearly impossible to ride. The pommel horse doesn't buck you off so much as it defies you to fly around on it at breakneck speed as it patiently awaits your basal ganglia, (home of your reflexive System 1's pre-programmed pommel horse routine), to relinquish the reins to your System 2's prefrontal cortex, (home of your rational second-guessing analytical self.)

A lack of confidence and insufficient myelination are the precursors to the unwanted hand-off from automation to awareness. During performance, as soon as you start *thinking* about what you're doing, you interfere with automation and problems arise. (This is akin to the computer's downloading process when the screen message warns you not to leave the screen or perform any other task during the download.)

There can be as much as a half-second delay for the sync up between your conscious and unconscious brain regions. A half-second is an eternity for neurotransmitters, more than enough time for a momentary stutter to cause you to lose your focus and screw up.

For the beleaguered gymnasts, once their analytical System 2 prefrontal cortexes grabbed ahold of their routines, they were forced to kiss the gold medal goodbye. When you start *to think* about what you're doing, and in essence wrestle the routine away from your preprogrammed unconscious, your routine splinters into a thousand pieces.

The 10,000-hour Rule

In the heart of the Sunshine State, Florida University professor of psychology Anders Ericsson has quietly spent the last 30 plus years fine-tuning his understanding of how deliberate practice works.

Malcolm Gladwell brought some of this work to light in his 2008 book *Outliers*,[161] writing about what he called "the 10,000 hour rule". However, in Ericsson's book, *Peak: Secrets from the New Science of Expertise,* Ericsson argues that the truth behind effective progress is much more complicated, and certainly more than just a number.[162]

We caught up with Anders Ericsson on a warm August day in 2016 outside of the University of Pennsylvania.

Our story begins in the fall of 1987, when Anders Ericsson took a position at the Max Planck Institute in Berlin, Germany. He was on a professional mission to determine the connection between motivation and expertise, or how practice affects domain mastery. Luckily for him, not far from the Institute was the renowned Berlin University of the Arts. The University's College of Music was turning out world-class musicians, so Ericsson figured he'd found the perfect place to conduct his experiment.

He asked the university officials to identify the violinists they believed had the potential to become international soloists, the future superstars of the violin world.

Ericsson and his research team decided to establish three groups of students, identified and ranked by the officials on their violin proficiency. The 'good' group was comprised of violinists who, although talented enough to be admitted into the College of Music, were still deemed more likely to end up teaching violin as opposed to performing professionally. The 'better' group was composed of violinists who had demonstrated unique talent, but were not likely to reach the caliber of play necessary to attain international soloist status. The final group was the 'best' group, who the University professors believed were the most likely to include future violin superstars. This group was thought to clearly be several notches above the good and better groups.

Each group included 10 students matched for sex and age to allow for consistency when interpreting experiment results. And in addition to the three groups, Ericsson and his team created a control group of ten middle-aged violinists recruited from two well-known Berlin orchestras with international reputations. The best group was deemed likely to end up playing in orchestras of this caliber, so the professional musicians represented a kind of future state for the experiment.

In 1987, the prevailing thought was that the difference in ability between all groups could be traced to individual innate talent, the genetic lottery winners being the future superstars of the violin world.

Ericsson's team did extensive background research on all experiment participants: when they started playing violin, who their music teachers had been, what their training program had been like, how many weeks they'd spent in solitary practice as opposed to competitions, their opinion on what drove their performance improvement, how often they took lessons, how often they'd given lessons, how much violin music they listened to, their performance experience, what level of music theory they had and how they'd acquired it, how much effort they expended in practice, how much pleasure they derived from playing versus practice, how many hours per week they spent practicing solitarily over the last ten years of their lives, and so on.[163]

As part of the experiment, the students in the three groups were also asked to keep highly detailed diaries of their daily activities and practice regimen over a period of a week.

When the experiment concluded, Ericsson's team noted some key takeaways. First, there was tremendous consistency among all groups. For instance, all groups agreed that solitary practice was hugely important for improvement, that intense practice was not enjoyable, and that taking regular naps and getting a good night's sleep was critical for spurring regular improvement. None of these results is all that surprising. The violin is an instrument that's been played for hundreds of years; practice and training methodologies are well established.

What was surprising—and what put Ericsson's research on the map— was the correlation between the number of practice hours and the ranking of each group. From the age of eight until their admittance into the College of Music at eighteen, the members of the good group had practiced a reported average of 3,420 hours, the better group had practiced for an average of 5,301 hours each, and the best group had logged an average of 7,410 hours of practice.[165] (The professional violinists from the two Berlin Symphonies had reported roughly the same number of practice hours as the best group over the same ten-year period in their own youth.)

The bottom line: the violin students' futures as international superstars were not predicated on their status as prodigies, but based firmly on the number of hours they'd put in while growing up. The study concluded that more practice was strongly correlated with future violin superstar status. In other words, more myelination means better skill acquisition.

Since 1987, these same findings have been repeated again and again, across a wide variety of domains. It seems that the right type and duration of practice is one of the biggest success factors, regardless of early talent.

This, Ericsson writes, is where writer Malcolm Gladwell went a bit off course.[166] Ericsson had mentioned 10,000 hours as the average amount of practice time the 'best' student violinists had put in by the time they were 20 years old. Gladwell interpreted this to mean that everyone in the 'best' group had practiced 10,000 hours. In fact, it was only an average; some practiced far more, some less.

To complicate this, Ericsson's original estimation was based on a fixed age. Had he chosen to use the number of practice hours acquired when the students were, say, 18 years old instead of 20, the 10,000-hour rule might have been the 7,410-hour rule. Ericsson makes a point of saying that although the 'best group' was highly accomplished, it would be a mistake to say that at 20, they had 'mastered the violin, even for those with 10,000 hours of practice under their belts.

In *Outliers*, Gladwell went on to extrapolate that it took roughly 10,000 hours of deliberate practice to be considered a master in many different domains. He posited that from Bill Gates to the Beatles, 10,000 hours was the magic number.

Gladwell noted that the Beatles clocked countless hours perfecting (and in effect, practicing) their songs in pre-dawn sets in Hamburg dive bars like the Cave. He theorized this is how they built their musical bona fides to become one of the most successful bands of all time. In musical parlance, this is known as paying your dues.

There is a problem with this reasoning, aside from the 10,000 hours problem. In his 2013 biography of the Beatles *Tune In,* author Mark Lewisohn chronicled their Germany period and calculated that the number of practice hours was closer to 1100—significantly shy of Gladwell's maxim.[167]

Lewisohn goes on to suggest that the true magic of the Beatles had far less to do with their playing abilities, and far more to do with the powerful songwriting duo of Lennon and McCartney. The real money might be in examining the deliberate practice methodology they employed when writing some of rock's most cherished canon.

Performing together night after night can certainly make for a tighter band. However, Ericsson's research suggests that true deliberate practice is a very structured regimen of setting incremental goals, constantly assessing progress through feedback (generally with the help of a well-seasoned coach), and making adjustments to better align with goals. Furthermore, this kind of practice requires intense concentration, and often a snail-like pace, in order to guarantee the building of flawless neural code. Finally, due to its focus on the very fine, specific points of technique, deliberate practice is generally done alone—and certainly not in front of an audience. Essentially, the Beatles were hard at work improving their chops, but their early years in German bars did not strictly adhere to the tenets of deliberate practice.

In a radio interview regarding Gladwell's 10,000-hour rule, former Beatle Paul McCartney said:

> "I think in our case, we always said, "Man, we had so much practice that by the time we got famous, we really knew what we were doing, and we were a good cohesive unit as a band,... I mean there are a lot of bands that were out in Hamburg, who put in 10,000 hours and didn't make it so it's not a cast-iron theory."

McCartney continued,

> "I think, however, when you look at a group who has been successful—you always will find that amount of work in the background. But I don't think it's a rule that if you do that amount of work you're going to be as successful as the Beatles."[168]

Gladwell has seemed to walk back the exactitude of the 10,000-rule a bit. In a later *New Yorker* article, Gladwell quotes a passage from an *American Scientist* article reporting on Nobel Prize Laureate Herbert Simon and William Chase's research regarding chess:

> "There are no instant experts in chess—certainly no instant master or grandmasters. There appears not to be on record any case (including Bobby Fischer) where a person reached grandmaster level with less than about a decade's intense preoccupation with the game. We would estimate, very roughly, that a master has spent perhaps 10-50,000 hours staring at chess positions..."[169]

In the same article, Gladwell tells the story of psychologist John R. Hayes, "who looked at seventy-six famous classical composers and found that, in almost every case, those composers did not create their greatest work until they had been composing for at least ten years."

Gladwell himself concedes, "I think it is also a mistake to assume that the ten thousand hour idea applies to every domain." He writes:

> "The point of Simon and Chase's paper years ago was that cognitively complex activities take many years to master because they require that a very long list of situations and possibilities and scenarios be experienced and processed. There is a reason that the Beatles didn't give us The White Album when they were teenagers."

Gladwell concludes by saying, "What Simon and Chase wrote forty years ago remains true today. In cognitively demanding fields, there are no naturals."

In this respect at least, Anders Ericsson agrees. Domain expertise requires a significant investment in dedicated practice, and there is no free pass, even for the so-called genius or prodigy. It's precisely this point that Ericsson has spent much of his scientific career proving.

Still, he says, although deliberate practice can generate impressive results, there are certain domains, like professional sports, where physical size precludes some individuals from a reasonable shot at commercial success. It doesn't mean that a person can't develop expertise in that sport, but that might not be enough to compete in the NBA or NFL.

Professional athletes aside, as Paul McCartney points out, there are plenty of people walking around who have devoted lots of hours to their passion, hobby or long term goal who have not acquired what we would call mastery over their chosen domain.

Why is this?

Reflexive System 1 at Work

Let's return to the scientist Daniel Kahneman's description of your brain's two major decision-making processes: System 1 (reflexive and emotional) and System 2 (analytical and methodical).

The beauty of your reflexive System 1 process is that it's always looking to preserve your precious mental fuel (glucose) for a later emergency.

The brain attaches emotional meaning to certain events. Collected together, these meanings form opinions, opinions create beliefs, and beliefs lead you to create rules for making decisions that you can then apply to new situations. In the scientific community, these rules are sometimes known as *heuristics*. Your brain catalogs your heuristics, using them for lightning-fast System 1 decisions. This saves you from having to consciously invent new guidelines for every situation in which you find yourself.

In other words, your brain conserves energy by reapplying rules that you've used in the past. This works pretty well most of the time.

Basically, your brain is gambling that a variety of life situations are similar enough that 'a one size fits all' approach to decisions will get the job done.

Our brains rely on—and would be lost without—our shortcuts, our network of belief circuits to help maneuver us through our day. Unfortunately, this kind of system biases us toward simplicity. We are more comfortable with black-and-white answers, often choosing not to examine the nuance of a process, decision or argument that might put our belief at risk.

It takes more mental energy and your analytical System 2 reasoning process to grapple with the gray area of a decision or argument, or to break down a practice regimen for exhaustive analysis. For many of us caught in the swirl of our day, relying on preconceived heuristics just saves time and energy. Who has time to spend digging into the reasoning, or lack thereof, behind the myriad of daily decisions?

Who wants to constantly examine the nuances of each stage while practicing a particular skill? When we're going through a presentation, rehearsing a song on the piano, or learning how to swing a golf club, we tend to focus on successfully making it through the process. That feels like the goal, to get from point A to point B, not to dissect and examine each and every incremental step of the journey along the way. And so System 1 tends to engage in shortcuts, seeking the fastest way to get from point A to point B.

The only way to teach your brain how to engage in deliberate practice is to actively question and examine your own performance heuristics. That can be a messy business, which can lead to uncertainty—one of the states your reflexive System 1 process is designed to avoid. There is reassuring safety in locking down on a heuristic and adamantly refusing to open it up for reassessment. Of course, a little uncertainty isn't necessarily bad. It might even make perfect sense, provided you've taken the time to work your way through your network of preconceptions and nuances that drive an argument or decision.

Your Broccoli Brain

Our mental shortcuts begin to form very early. Imagine your beleaguered mom trying to convince three-year-old you to eat broccoli for the first time. With a little maternal coaching, you dutifully take your first bite and reflexively spit it out. Your assessment: it's not sweet or salty and doesn't have a particularly good mouth feel.

Your mom employs all the standard mom tricks to entice you into eating broccoli, including describing the broccoli as little green trees from a tiny forest, and wouldn't you like to eat a miniature tree? You, of course, are not so easily conned by the forest gambit and so you hold steadfast to your refusal. Your mom has had a long day and she's tired, so eventually, albeit reluctantly, she gives up. She, of course, is not without a plan, because she's already relying on a much better rested Future-Self to win the broccoli battle in the next round.

Sure enough, a couple of weeks go by and your mom takes you on a second time with a steaming plate of broccoli. However, you maintain your position. By now you've realized that, short of an intravenous broccoli tube, it's just a matter of holding out a little longer, and the victory is yours. If we could peak inside your three-year-old skull at this point, we would see that you are starting to wire up a little brain app concerning broccoli. Your new heuristic can be summed up rather succinctly: "I don't eat broccoli."

At three years old, you've created your broccoli rule and now that it's in place, you don't have to think about it anymore. At the mere mention of broccoli your brain fires up your anti-broccoli circuitry.

Fast forward: you're twenty-one and your new girlfriend or boyfriend invites you over for a home-cooked meal. Taking your seat at the table, you notice a familiar side dish: a freshly steamed bowl of what resembles little green trees. Without a second thought, you mutter, "I don't eat broccoli." Boom, it's that simple—no deliberation.

Never mind that broccoli is a food rich in antioxidants, and generally one of the healthiest foods you can eat. Your rule is firmly in place; it's operated unimpeded for eighteen years. The problem with this heuristic is that you built it when you were three years old. At three, broccoli seemed unthinkable. Compared to any multicolored sugary breakfast cereal, your culinary gold standard, broccoli wasn't even worthy of an afterthought.

Now, at twenty-one, you haven't considered going back and updating your old rule. Bottom line, you're still relying on the judgment of a three-year-old for your nutritional advice. Does that make sense? Imagine that a really precocious three-year-old sidles up to you and offers you some stock tips. As an adult, would you take stock advice, or any kind of serious advice, from a preschooler? It's unlikely you would.

We rely on our reflexive System 1 shortcut rules all of our lives, regardless of the time, place and conditions in which those rules were established. Surprisingly, for the most part, that works out pretty well. After all, there are about seven billion of us on the planet and counting. As a species, we've managed to do quite nicely with our current brain configuration.

Chances are, when you get to a busy crosswalk, you stop and wait for the light to change and traffic to clear before you walk. That's a pretty sound rule. You've been following that heuristic since a very young age.

You don't waste your System 2's precious analytical resources calculating the traveling speed of the oncoming delivery truck to see if it's worth gambling on a mad dash across the street while the light is still blinking red.

Seldom do we employ our System 2 analytic powers to question the current value of System 1's pre-built heuristics. Our System 1 reflexive shortcuts eliminate the need for about half of the decisions we would have to consciously make during the day, freeing our System 2 processes to worry about and ponder plenty of other important decisions.

Tricking your Reflexive System 1 Process

Now we know that when it comes to practice, the Holy Grail of learning, reflexive System 1 is both your ally and your enemy. Performing the same actions over and over means activating the same neural circuitry. To expend less effort, System 1 insulates those circuits with myelin. The more myelin, the stronger and faster those signals—Hebb's rule in action.[170] When everything goes right, we call this skill.

Skill acquisition relies on repetition, but System 1, in its drive for efficiency and glucose conservation, is quickly satisfied. System 1 is, and this is really important, willing to apply a new rule any time it recognizes repetition. It isn't interested in absolute skill mastery or perfection. Refinement is not its concern. Your reflexive System 1 motto might best be described as "If it works okay, good enough."

So given System 1's preference for quick fixes over quality, how can we move forward toward flawless skill mastery?

In many ways, this is the fork in the road where the dedicated leave the rest behind. For the serious, the *pit of practice* is where mastery is forged, through hard, deliberate work. This is where we might argue 'genius' is truly built, brick by brick.

Keys to Deliberate Practice: Time to Kaizen

Masaaki Imai, in his pivotal book on competitive success called *Kaizen,* uses the word "Kaizen" to describe the idea of continuous improvement.[171] The concept involves seeking out small, incremental, positive shifts in a given process, and benchmarking them as you go, always with an eye toward perfection. In business, it means all hands on deck for everyone from top management to the worker on the floor. It's everyone's job to take an active role in searching for ways to improve the quality process.

Kaizen requires combining reflexive System 1 and System 2 analytical processes during both the learning and practice phase. Although the goal of absolute perfection is never achievable, the sheer act of relentlessly pushing to be just a little bit better makes stagnation unlikely.

Imai explains that the theory of continuous improvement was the winning strategy that allowed the Japanese car industry to gain a strong foothold in the U.S. market.

At one time, Japanese automotive quality was subject to extensive ridicule and anemic U.S. sales, until they changed their manufacturing model. Ironically, those changes can be traced, in large part, back to American W. Edwards Deming, considered by many to be the father of modern manufacturing. Particularly in Japan, he is hailed as a hero.

After World War II, Deming traveled to the devastated country as a statistician to help rebuild the economy. There, he developed his 14 key principles.[172] These 14 commandments of modern manufacturing helped turn Japan into the global powerhouse it is today. Central to Deming's philosophy is the notion that, when solving a problem, it is first necessary to identify the root cause, and then engage in continuous improvement until the problem is eradicated. This is the basic operating principle behind Kaizen.

On the surface, it doesn't seem too groundbreaking. (But then again, neither does washing your hands before surgery, and that only caught on in the twentieth century.[173]) Failing to follow Deming's root-cause rule has caused incalculable cost in manufacturing, medicine, politics and personal relationships, to name but a few.

In the Kaizen process, you have to be willing to tolerate a measure of uncertainty and accept the risk that some solutions will undoubtedly fail and require alternative solutions. Growth mindset is a central tenant of Kaizen. Trial and error is accepted as a necessity to achieve the end result. This tolerance of a certain amount of pain requires both patience and our old friend, willpower.

By uncompromisingly striving for perfection, the Japanese automakers gradually improved their quality and eventually caught the American car industry by surprise. By the early eighties, the Americans were borrowing Deming's 14 principles and the idea of Kaizen to improve their own processes in an attempt to catch up.

Staging your Own Kaizen Event

The Kaizen concept ties in nicely with Ericsson's idea of deliberate practice. Deliberate practice can be represented as the following five fundamental rules:

1. **Deconstruct.** Understand your current skill level and break the subject matter into smaller chunks, making sure the chosen practice techniques are just within your ability. Practice the chunks separately. Tackling something as a whole is overwhelming and can lead to frustration and failure.

2. **Mental Representations.** Build your domain knowledge as deeply as possible, for a thorough and specific understanding of the desirable goal and a reasonable idea of what it takes to achieve it.

3. **Quality repetition.** Your goal must be perfect neural code. Work on each chunk carefully, slowing down to guarantee you've fully mastered each section before moving on to the next one. The goal is incremental improvement; otherwise, you'll myelinate mistakes right into your repetition and neural code. There is no shortcut around the amount of quality repetitions needed to gain mastery.

4. **Feedback Loop.** Find a domain expert (coach) who can observe you in real time and give you immediate feedback about what you're doing right and wrong. Then demonstrate a growth mindset: be open to critique and the incremental steps you can take to improve performance. It's the coach's job to ensure the practice environment has enough creative variation to keep practice dynamic. Coaches with true domain knowledge use their own mental representations to help facilitate your deeper understanding.

5. **Rage to Master.** Push yourself to the edge of your ability. Growth generally involves a little mental, and sometimes physical, pain. The rage to master means you are significantly motivated to hang in there for the long haul, no matter what obstacles may come your way. Seek to simulate a practice environment where your practice intensity and performance intensity are one and the same.

It should be noted that the rage to master is not a principle that Ericsson cites in his work on deliberate practice.

When we asked him about the idea of rage to master, he told us he didn't see it as a free-standing trait. Rather, it was domain specific and implicitly embedded in the activity, intensified during practice through steadily increasing, higher quality mental representations, and facilitated through feedback loops and coaching. He suggested the process itself created drive, a kind of internal momentum. He explained the difference between musical masters and average music students in this way:

> "[What] seems to distinguish people who eventually become successful musicians, is that internalization where they are actually into making music as opposed to the traditional music student who's just learning how to do things.

> "If you were to tell [traditional students] to make changes to what it sounds like, they don't have a clue because they can't really hear what they're playing. And I think that's true in so many other domains, [mental] representation allows you to start having internalized control over how you actually get better and how the training is linked into you being able to do things you couldn't do before.

> "What we find is—not engaging in these kinds of deliberate activities where you actually get feedback of whether you're right or not—performance is actually pretty unimpressive."[174]

Rule 1. Deconstruct

Deliberate practice begins with identifying the smallest parts that make up the whole of a desired skill. The idea of incremental improvement, i.e. perfection of each small step generating a large consequence, is best illustrated by 'force amplification,' or what is more commonly known as the 'domino effect.' In 1983 Physicist Lorne Whitehead, of the University of British Columbia, demonstrated the enormity of this concept.[175]

Imagine a domino, five millimeters high, a little smaller in size than a TicTac candy. Now imagine 28 more dominoes each 1.5 times larger in size and weight than the next one—all lined up ready to fall. Although you could literally knock down the first TicTac sized domino with a toothpick, the energy transfer magnified by the 29th domino would be enough to theoretically knock over the Empire State Building.

The point is that we often dismiss the powerful effect in setting off a chain reaction of small but meaningful incremental improvements.

It's a common human bias to swing for the fences and then give up when we don't achieve the desired result in the short run. Yet legendary baseball Hall of Famer Ty Cobb built his .366 lifetime batting average on the back of singles and doubles, not swinging for home runs.[176]

Practicing deliberately is really about thinking incrementally and building a new skill step by step. So the first step is vital to the process. Someone had to lay that first stone at the pyramid of Giza, Itzak Perlman had to run his bow across a violin string for the first time, and Serena Williams had to make her first serve. As the process plays out, tiny steps build into something much greater than merely the sum of parts. Deconstructing a process into its smallest pieces allows for a level of focus and mastery that can pay huge dividends when all the increments are eventually reassembled.

Research and Development

In the years leading up to World War II, the Marx Brothers were the toast of Hollywood. Their zany antics, ribald double entendre, and disregard for authority proved to be a winning comic formula, not just for their day, but for generations to come. Watching Groucho, Chico, and Harpo's now-classic performances, it's tempting to believe that they must have been endowed with the magical comic gene, winners in the cosmic humor lottery.

With the exception of Zeppo, the youngest brother, who eventually tired of being shoehorned into straight man roles and quit the movie business, the Marx Brothers showed a near-maniacal dedication to perfecting every nuance of their material. They had spent years in Vaudeville honing their craft, practicing relentlessly to hit the perfect comedy pitch in front of raucous and often unforgiving audiences.

When they finally got a chance to bring their wacky style of comedy to the motion picture industry, they weren't about to take any chances. Scenes, which seemed on screen to be moments of absolute improvisational genius, were in reality painstakingly blocked out, endlessly rehearsed and memorized.

To that end, they broke movie scripts down into smaller sections and went on the road, testing out bits for several months at a time before filming. Performing before live audiences to work out the unfunny kinks gave them an idea, in real time, whether the material had legs.

They even sometimes employed former Vaudeville actors to perform their own material in front of them for a more objective opinion on script quality.

Groucho was said to be the most obsessive perfectionist of all the brothers.[177] A Marx Brothers publicist, Teet Carle, recalls a scene in *A Day at the Races* when Chico is trying to sell Groucho a discounted book:

CHICO: One dollar and you remember me all your life.

GROUCHO: That's the most nauseating proposition I ever heard.

Carle reports that Groucho was fixated on finding the perfect punch line, cycling through numerous substitutions for the word *nauseating* which included *obnoxious, revolting, disgusting, offensive, repulsive, disagreeable, and distasteful*. For some reason, in the end, none of these words proved to be as funny as "nauseating".[178] The fact that we're still watching Marx Brothers movies today suggests they were on to something.

As hall of fame Green Bay Packer football coach Vince Lombardi once said, "Practice doesn't make perfect, only perfect practice makes perfect."

Long before the Marx Brothers hit the stage and screen, a young writer in Missouri was struggling to perfect his own wordplay. Samuel Clemens, a.k.a. Mark Twain—printer, steamboat pilot, journalist, failed miner, failed entrepreneur, political commentator, humorist, and lecturer—is considered perhaps to be the greatest of American storytellers. Ernest Hemingway once declared, "All modern American literature comes from one book by Mark Twain called *Huckleberry Finn*...It's the best book we've had."[179]

At first glance, Twain's beginnings may seem to make him an unlikely literary hero. The man whose books are assigned to every high schooler in America only finished fifth grade. When Twain was twelve, he lost his father and left school to apprentice at a local printing shop in order to help with the family expenses.

At fifteen, he was working as a typesetter in his brother Orion's newspaper office. Setting type in those days meant assembling each word one at time, letter by letter. It was a painstakingly slow process, but one that allowed an aspiring writer to develop a deep domain knowledge of just how the English language went together, down to the clauses and commas. This might be where Twain came to understand the power of language as exemplified by his famous proclamation, "The difference between the almost right word and the right word is really a large matter—'tis the difference between the lightning-bug and the lightning."[180]

At seventeen, Clemens set off from Missouri to New York City, seeking fame and fortune. He began writing for magazines and newspapers at twenty-two. It wasn't until 1869 at the age of thirty-three, that his first book, *Innocents Abroad*,[181] was finally published. By this time, he already had eleven years of writing under his belt.

Twain was no stranger to a work ethic of practice. In a letter to his good friend, editor and literary critic Dean Howells, Twain reflected on the writing of what would become *The Adventures of Huckleberry Finn:* "I wrote 4000 words today & I touch 3000 and upwards pretty often, and don't fall below 2600 on any working day."[182] The thought of a legend like Mark Twain measuring his progress in such a dry, methodical way may take a little of the romance out of writing the great American novel. Still, there is no escaping the fact that to be a writer, one must do a tremendous amount of writing. It's only through a long process of constructing, deconstructing and then reconstructing through edits that a final paragraph can look effortless to a reader.

In the evening, part of Twain's writing ritual was to gather his wife and daughters around him and read through what he'd written during the day. This was a vital part of his process, explains biographer Justin Kaplan, because Twain was "chronically incapable of self-criticism; he relied on others to make the basic judgments of his work."[183]

Twain's wife Livy was born into a wealthy, progressive eastern seaboard family and received an extensive formal education. Twain was the lucky beneficiary of Livy's classical knowledge. His family served as his editorial board, listening and making suggestions. Twain acknowledged that he seldom overrode their input, and that he relied on their judgment of when his work was up to snuff, and all the parts and pieces fit together.

The Jigsaw Puzzle Dilemma

Have you ever encountered a jigsaw puzzle that happened to be missing a few pieces? Of course, you can't realize this until the puzzle is near completion. Upon discovery, it's very disconcerting— you put in all that effort, but now the picture will never fully come together. Interestingly, from a strict percentage point of view, if two pieces are missing on a 500-piece puzzle, you still have over 99 percent of the set. Still, chances are good you will simply throw the whole puzzle away. Why? Our brains love the idea of a complete story. We're hardwired for it.

It is easy to stand in awe of great accomplishments and discount the effort behind the scenes to make it happen. When we encounter a masterful comedian improvising, a ballet dancer leaping through the air, or the Swampers' session work, we perceive only the final product. Our brain understands that demonstration of skill as a unified moment, without context. What we don't see are the countless hours of effort that created that moment, the long march of minuscule advancements, achieved through grueling and dedicated practice. Simply put, we see the picture on the cover of the box, not the slow careful assembly of hundreds of puzzle pieces.

Peeling back the curtain tells us that there is a constant among the masters, and that constant is adherence to the rules of deliberate practice. Learning is best achieved by breaking a task into the smallest increments possible and working on each increment until you've mastered it. Twain did it letter-by-letter, word-by-word, and sentence-by-sentence.

It takes real work and a lot of glucose to stop and mentally examine each of our minute actions. However, that careful analysis on the front end pays off in the long run.

Dave Guy has spent many hours practicing his trumpet. Even today, he from time to time pulls out his old music manuals from school and brushes up on the basics. But what struck us the most in speaking with Dave was the way he described his practice. He made it clear that for him, it isn't simply about banking hour after hour of blowing scales on his horn. What matters is the musical tone, and the emotional connection he feels when he plays.

Dave is relentless about producing a sound. Put simply, Dave's practice is organized around quality, not quantity. Sure, over the course of his career he's logged his hours, but those hours were in service to his sound, not a rote schedule. For Dave, the sound comes before anything else. That's part of what makes his music so compelling to the listener.

Rule 2. Mental Representations

In 1501, the Church of Florence commissioned Michelangelo to sculpt a marble statue of David, the Biblical shepherd boy. Because the statue was to be placed on a hill and viewed from below, it needed to be about thirteen and a half feet tall—roughly twice life-size.

Proportionally, this created some challenges, but to make it an even greater test of ingenuity, the marble Michelangelo had to work with was not pristine.[184] It was a leftover from an unfinished 1464 sculpture by another artist.

Clearly, Michelangelo had his work cut out for him. At this point, the artist was only in his twenties, but he'd already served his time in apprenticeship and made a name for himself with his *Pieta* sculpture in Rome. More to the point, he had spent so much time and concentration honing his skills under masterful supervision, that even working with another sculptor's scrap, he was able to free the David from his marble prison.

At first read, this might sound like a nice metaphor about art emerging from nature, but this envisioning is at the critical mass of his successful sculpture. According to Ericsson, what sets experts apart from their competition is not just raw talent or the practice hours logged; it's what Ericsson calls "mental representation." Ericsson writes that true experts have developed the ability to dream up an idea down to the most minute details and reproduce it in the real world exactly as envisioned.[185]

When master artists come face to face with a marble block, their goal isn't just to carve out a generalized form; they are able to aim for an incredibly specific image—say, a shepherd boy waiting to vanquish his oversized rival.

Throughout my travels, I've had the opportunity to conduct some informal sculpting experiments of my own. Sometimes, I will give the adults in my seminars ten minutes, some play-dough, and instructions to make a fist-sized elephant. None of my subjects have any formal training in the arts. They generally begin by producing the necessary elephant parts, which include four legs, a tail, a trunk, tusks, floppy ears, and so on. Once the parts are formed, they tend to assemble their sculpture like a puzzle. The result is a crude, albeit recognizable, rendering of an elephant.

However, giving them more time does not tend to produce a better elephant. The deadline is not the issue. It seems they either lack an internal picture of what an elephant truly looks like, or the hand-eye coordination to translate that picture to play-dough—or both.

Ericsson says that hyper-nuanced specificity, a note-for-note model, of the exact goal is the X factor in deliberate practice. People who successfully perform the type of practice that leads to real improvement start with a crystalline vision of the end product. Like Michelangelo, they begin work already seeing the David in the stone, or the elephant in the play-dough. It's not some stylized, cartoonish notion of body parts; they envision everything down to the veins in the bicep or the bark-like skin of the elephant's hindquarters.

Once you can conceptualize your end result, you can develop a systematic strategy for getting there, performing the steps necessary to bring it to fruition.

The problem for novices is that they often haven't acquired the domain knowledge to imagine that perfect prototype yet. Do I really understand precisely how this Beethoven sonata should sound, or how the baseball should appear as it leaves the pitcher's hand? Ericsson's research suggests that the novices who do possess better mental representations tend to produce faster and more accurate work.

For instance, Ericsson posits that mental representation allows for a more refined level of pattern recognition. It's not that a professional baseball player's reflexes are necessarily faster than yours or mine, but the knowledge base to recognize a subtle change in the pitcher's wrist that prepares the hitter to move into a better hitting position.

This acute sense of anticipation is the winning formula for everything from chess to playing the piano. It's the essential hack for skill. True effective practice is not just about performing the same rote actions again and again; the real pay off is developing a deeper understanding of the domain-specific mechanics.

Ericsson says that both the quarterback spending countless hours watching game films and the chess player endlessly studying past grandmaster matches, are wiring a vast library of patterns that their brains can then run like unconscious algorithms when competition heats up.

The lack of a wholly accurate, detailed mental representation is a little like having a general idea how to get from the east coast to the west coast, but not knowing your destination city, let alone the street address. And a hazily defined goal or incomplete pattern recognition leads to mistakes, frustration, and frequently, some very odd-looking play-dough elephants.

Rule 3. Quality Repetitions

Repetition on its own does not guarantee perfection. One can practice any kind of routine over and over again, but if the sequence contains flaws—a warble in your voice, an awkward bit in your sales presentation— then your hours of practice are likely to be just that, hours of practice.

Your brain encodes the information you give it. As they say in computer programming, "garbage in, garbage out." Flaws and glitches in computer programs cost businesses millions of dollars in loss every year. When you're building a brain app, flaws and glitches in your practice regimen relegate you to *stagnation swamp.*

Deliberate practice requires not just time, but growth. Your reflexive System 1 is perfectly happy to engage in consistent practice, but by its energy-saving nature, you can't count on System 1 to push for growth and perfection. This part of your brain is about efficiency above all else, automating whatever you learn, regardless of whether you've achieved mastery.

Computer software companies consistently offer up new and improved versions in an attempt to fix the bugs in their current software. They have teams of people reworking the code. When it comes to building intentional neural code, you are most likely an army of one.

This is why, thinking back on unfulfilled goals, you might have made rapid progress at first (logarithmic growth), then progress slowed and leveled off, with a couple of rough patches in place that never got totally ironed out.

System 1 automates what you give it; quality control is another department. That department is System 2, the effortful part of your brain that can analyze and critique your actions. When it comes to learning, skilled masters are continuously studying and improving their own results. They know better than to settle for the instinctive "good enough" of System 1.

Ericsson says that it's a mistake to simply equate deliberate practice with System 1's rote repetition alone. He told us:

> "I think, sometimes people point to deliberate practice as this 'drill and kill' [idea], that nothing is going on in people's heads, when in fact, when I talk to elite performers, they have some very specific ideas of what they want to change during training."

Problem solving is essential to working out the kinks in acquiring any skill.

Prepare for Problems

When it comes time for the actual in-the-moment performance of a skill, it's unlikely that conditions will be perfect.

On July 19th 1969, as the world looked on, Apollo 11's Eagle landing craft rapidly descended toward the surface of the moon. Gauging against landmarks on the terrain, the astronauts quickly realized they were about four seconds ahead of their target, a distance of several miles off their intended landing site. A glitch in the computer landing software had pushed the craft off course, and Commander Neil Armstrong could see they were headed for danger, an area strewn with large boulders—potentially a fatal landing.

Thanks to hundreds of hours in a flight simulator back on earth and ample experience practicing against as many scenarios as NASA scientists could invent, Armstrong coolly switched over from automatic mode to take control of the craft and, with the help of his fellow astronaut Buzz Aldrin, successfully guided their craft to a safe landing. When Apollo 11 finally touched down, they only had 25 seconds of fuel left—not a lot of margin for error.

Without systematically deconstructing your game plan or practice patterns to look for potential weakness, it's easy to overlook a factor that might swamp your efforts. Relying too much on routine can lull you into a false sense of security. Had Armstrong and Aldrin not previously run the gauntlet of possible problems, the landing might have ended without "One giant leap for mankind."[187]

You can see how a fixed mindset can hurt you. If you are afraid of even momentary failure, you are more likely to avoid the nerve-wracking business of anticipating every possible setback and practicing to meet a wide variety of challenges. Changing the way you've learned to do something, once you've memorized a pattern, feels like an invitation to fail, rather than a chance to improve or succeed.

Engaging in a growth mindset during practice requires continuous evaluation by your System 2 process, staying vigilant to even the smallest of potential mistakes. Then the trick is to employ a mini-Kaizen event to guarantee quality neural code.

Six Steps toward Quality Neural Code

1. Ask yourself, what is causing the glitch or problem?

2. Deconstruct the process and identify all the pieces and parts that make up the whole.

3. Create a checklist to guarantee you haven't missed anything.

4. Have a domain expert (coach) verify your plan. With the help of your coach craft a deliberate practice solution to address each piece or part separately.

5. Practice in concentrated, highly focused 15-minute sessions.

   a. Make sure you've gained absolute qualitative mastery over a particular piece or part before moving on to the next section.

   b. Practice the final production (the aggregation of all the pieces and parts) under real life conditions until it's second nature.

6. Have a domain expert regularly review, tweak, and help you vary your practice regimen. It's important for the domain expert to judge the final product or outcome of your work.

We've seen in growth mindset that learning from failure is one of the keys to mastery. This is part of why practicing in front of a live audience can be invaluable—they may spot points of possible improvement you wouldn't recognize on your own.

Rule 4. Feedback Loop

Swimmer Michael Phelps is the most decorated Olympic athlete of all time. Of his 28 Olympic medals, 23 are gold.[188] Phelps trained in Ann Arbor, Michigan, under swimming guru Bob Bowman. Swimming is a highly repetitive sport, involving long hours in the water, perfecting the efficiency of every stroke. The tedium of doing laps can be meditative, but over-reliance on routine can lead to complacency.

One day in practice, Bob Bowman purposely stepped on Michael's goggles, ensuring they would leak.[189] A pair of cracked goggles would force Michael to cope with the sting of chlorine in his eyes and to operate with very limited vision.

Championship swimmers need to know exactly where their bodies are in relation to the pool wall; it's how they judge when to turn. A few microseconds lost on a badly timed turn can literally spell the difference between a gold medal and a consolation prize. Phelps would instead have to count, memorize, and feel the number of strokes from end to end—in essence, swimming blindly.

Practicing in intentionally extreme conditions is sometimes known as 'bookending'. Instead of preparing yourself to attack a goal under the best-case scenario, you purposely make the conditions harder than expected to give you a psychological edge. The thinking goes, 'If I can perform well under sub optimal conditions, I'll dominate under normal conditions.' This keeps you from falling into the dangerous trap of complacency.

In the 2008 Olympic games in Beijing, China, Coach Bowman's leaking goggle strategy paid off in a big way. During his 200-meter butterfly attempt, Phelps's goggles failed and began to fill with water. Phelps, of course, knew exactly what to do. He kept his focus and ended up setting a new world record.

This is part of how even a champion swimmer like Phelps continues to benefit from an insightful coach. If the true goal is complete mastery, a person with considerable skills still can't afford to rest on his or her laurels. It's the coach's job to keep you on the road of incremental improvement and keep you from plateauing. Bowman has been described as a 'mad scientist' constantly finding new unusual ways to keep his swimmers kicking through hours of grueling practice.[190]

Finding a coach who is a true expert in the domain of your choice can be a challenge. Ericsson cautions that there are plenty of individuals that purport to have expertise, so make sure you have some objective way of verifying their real knowledge level. A coach must have superior mental representations of the domain you are seeking to master, along with the 'playbook' necessary to get you there.

With proper coaching and established feedback loops, eventually you'll better recognize the quality of your performance through your own clearly defined mental representations.

As noted, one of the most important things a good coach can do is to emulate Bob Bowman's search for more creative and surprising practice techniques. Variety is vital to keeping practice fresh and keeping you advancing steadily toward your goal.

An accurate critique of your mechanics can lead to a new level of understanding. Unfortunately, as Michael Phelps can attest, sometimes, it stings a little.

Rule 5. Rage to Master

Coasting: it's a simple idea. For skateboarders, finding a long incline is a dream, the chance to relax and glide toward the bottom, picking up speed and momentum. However, in deliberate practice, coasting is a plateau. Growth is achieved through constantly challenging yourself, incrementally building on your last achievement. Rage to master is obsessively working toward the unattainable goal of perfection, never fully satisfied with your last achievement.

Rage to master is Roger Bannister pushing himself to break the four-minute mile barrier as his lungs burned and his legs told him to quit. It's Stevie Ray Vaughn practicing his blues guitar licks until he tore his fingertips, and then super gluing them so he could keep on playing.[191] It's young Mia Hamm deliberately seeking out faster, stronger soccer players and then forcing herself to keep up.

Or it's the inexhaustible Bruce Springsteen, dancing across the stage, still unleashing machine gun guitar riffs well into his sixties. Springsteen treats each concert as if he still has something to prove. Much like trumpeter Dave Guy, his burning desire to play goes back to grade school. And just like Guy, Springsteen was shy and not an early standout.

Although it would take decades for Springsteen to craft and hone his performance skills, those who know him well say his musical drive was undeniable from the first time he heard rock n' roll on the Ed Sullivan Show.

New York Times book critic Robert Ford writes that Springsteen had a "near-feral discipline—a studious and encyclopedic knowledge of the genre and rock history. An ungodly number of irreplaceable life hours spent practicing, practicing, practicing in small, ill-lit rooms." And, perhaps most importantly, "A ruthless calculation to be nothing less than great, powered by a conviction that greatness *can* exist and be redeeming."[192]

The good news is that rage to master, that inner drive and well of motivation, can be developed like any other skill.

Motivation seems to strengthen through incremental improvement: tiny wins build confidence. This happens when a part of your brain called the striatum forges a connection between your prefrontal cortex's analytical decision-making System 2, and the basal ganglia, the home of neural programming. In other words, the striatum acts like a switch connecting an idea and an action.

Linking idea and action gives you a sense of control over your efforts, which, research indicates, increases your willingness to repeat the action, especially when the action is successful. This creates a domino effect. The more control you feel, the greater your motivation, and the greater your motivation, the more you repeat the action. Engaging in deliberate practice is therefore an essential ingredient to developing your rage to master.

In his autobiography, Springsteen writes, "If you want to burn bright, hard *and* long–you will need to depend upon more than your initial instincts. You will need to develop some craft and a creative intelligence that will lead you *farther* when things get dicey."[193]

Faustian Bargain

In examining the lives of great achievers from T. S. Eliot to Martha Graham to Mohandas Gandhi, Harvard psychologist Howard Gardner makes a critical point. He suggests that expert level mastery often requires a certain Faustian bargain.[194] This means sacrificing other important areas in your life in order to harness your full focus and attention toward your passion.

"My study reveals that, in one way or another, each of the creators became embedded in some kind of a bargain, deal, or Faustian arrangement, executed as a means of ensuring the preservation of his or her unusual gifts," wrote Gardner."[195]

The partnership of sacrifice and mastery has been forever immortalized in the story of legendary bluesman Robert Johnson. It seems that Johnson did not begin as a prodigy–just the opposite. Famous Delta musicians like Son House claimed young Johnson had been embarrassingly bad, pestering anyone with a guitar to teach him how to play. During breaks in Son House's performances, he remembers Johnson's audacity of borrowing Son House's guitar. In an interview many years later, Son House recalled:

> "And such a racket you never heard! It'd make the people mad, you know. They'd come out and say, 'Why don't y'all go in and get that guitar away from that boy! He's running people crazy with it!' I'd come back in, and I'd scold him about it, 'Don't do that, Robert. You drive the people nuts. You can't play nothing.

Why don't you play that harmonica for'em.' But he didn't want to blow that. Still, he didn't care how I'd get after him about it. He'd do it anyway."[196]

In 1935, as the legend goes, Robert Johnson left town, wandered to a lonely crossroads at midnight, and sold his soul to the devil. When he resurfaced three months later, Johnson was the best blues player anyone had ever heard.[197]

The true story is probably much less spectacular, but no less interesting. Some historians suggest that Johnson left town for considerably longer than three months, and instead of the devil, he found a talented bluesman named Ike Zinnerman, who agreed to mentor the young would-be musician.

We can speculate that under the mentorship of a presumably highly skilled player like Zinnerman, he found himself in the perfect deliberate practice laboratory, building high quality mental representations through the feedback loop of tutelage and practice, culminating in performance. Combine this with his rage to master and—perhaps as importantly—the limited job prospects for a black man in the 1930's American south, and we might say he was doubly driven to succeed. Johnson spent the rest of his tragically short life playing and singing on street corners and juke joints. He never had a permanent address, traveling anywhere for a willing audience, and displayed little concern for how his lifestyle was perceived by others. Performing consumed his life.

Between 1936 and '37, Johnson recorded 29 songs for the American Record Corporation, including "Hellhound On My Trail," "Walking Blues," and "Sweet Home Chicago."[198] Johnson's self-imposed traveling schedule meant he eschewed close relationships, but unfortunately for him, he had an eye for the ladies. In 1938 it was said he was poisoned with tainted whiskey by a jealous husband while on a break during one of his performances. He died in relative obscurity at the age of 27.

Luckily, his musical legacy lives on. His small but influential body of work has inspired covers from a whole host of artists, including Scott

Ainslie, Bonnie Raitt, Elmore James, Eric Clapton, Bob Dylan, The Allman Brothers, and the Rolling Stones.[199]

Keith Richards, hearing Johnson's recordings for the first time, reportedly asked, "Who is the other guy playing with him?" "I was hearing two guitars, and it took a long time to actually realize he was doing it all by himself," Richards told an interviewer.[200] Eric Clapton is quoted as saying, "I have never found anything more deeply soulful than Robert Johnson."[201]

The good news is that a bargain with the Devil is not a prerequisite for pushing yourself to the edge of your abilities, but as Gardner and Anders Ericsson remind us, there is no getting around the fact that the time commitment and level of intensity and focus necessary to achieve expert level mastery creates a certain amount of sacrifice in your life.

In sports, there is the oft-cited 'No pain, no gain'. On some level, that might be a motto for the rage to master. Of course, outside the arena of athletics, that pain is for the most part mental–not just putting in the practice hours, but practicing every second with performance level intensity.

Maintaining a rage to master can be taxing. It pushes against our natural human behavior. This is where you must leverage growth mindset, **SMARTR** goal setting, and the exercise-nutrition-sleep triad, to help fuel your passion, your own rage to master.

And that's precisely why, while you can hack parts of the system, there is no one giant shortcut. Your real opportunity is building the necessary brain apps in the right sequence, which creates the greatest likelihood that you'll be able to hang in there till the final result. Luck aside, it's likely you'll need to harness your rage to master to make a successful run at your long-term goal.

Deliberate Practice in Action

From the outside, the brown brick SPICE building blends right in at the leafy green campus of Webster University in St. Louis, Missouri. Take a step inside and the giant chess board, populated by chess pieces the size of small children, signals that you're in for something different.

SPICE, or the Susan Polgar Institute for Chess Excellence, has many missions. SPICE aims to bring more attention to chess, to give all American schoolchildren access to the game, and to support women's chess. But one main goal is to create the next generation of chess geniuses for the Webster University chess team.

Founder Susan Polgar has first-hand experience on that count. We met with her on a rainy day in May 2015 at SPICE headquarters.[202]

Susan is the oldest child of Hungarian psychologist Laszlo Polgar. When Laszlo was a university student in Hungary during the sixties, he was fascinated with what made exceptional people tick. At one point early in his career, he says he examined the biographies of 400 of the greatest thinkers, across a wide range of fields, everyone from Socrates to Einstein. Two commonalities emerged: they all started their quest for expertise very young, and they all studied their work in a very specific and deliberate way.

"A genius is not born but is educated and trained," he would later tell a Washington Post reporter. "When a child is born healthy, it is a potential genius."[203]

In 1965, then nineteen-year-old Laszlo began a courtship of sorts with a young Ukrainian foreign language teacher named Klara. Fiscal resources, transportation and phones being what they were in the Soviet block, the relationship was conducted by mail. As the correspondence became more intense, Polgar asked Klara if she was willing to commit to a very long-term experiment: if the two were married, would she help Laszlo raise some geniuses of their own? Klara agreed.

By the time they'd married, Polgar had outlined his theories in a book called *Bring up Genius!*[204] In 1970, their first child, Susan, was born, and it was time to put those ideas to the test.

The Polgars took the radical step back in those days of homeschooling their children. The Hungarian school system took great umbrage, but Laszlo's theory depended on having strict control over the learning environment. The question was, what field should little Susan tackle? Laszlo and Klara had considered higher-level mathematics and possibly foreign languages, but ultimately it all came down to a twist of fate and the whims of a child.

One day, Susan was rummaging around their apartment in search of toys when she discovered an abandoned chessboard and chessmen crammed behind a cabinet door. She asked her mother about the little pieces. Klara explained it was part of a game, but Susan would have to wait for her father to get home because Klara didn't play.

"Yes, he could have put us in any field," Susan would recall later, "but it was I who chose chess as a four-year old...I liked the chessmen: they were toys for me." As with Dave Guy's decision to invest his time in the trumpet, Susan's first step toward virtuosity was an impulse rather than any long-term strategy.

Laszlo enjoyed the game, but he'd only ever been a casual player. This did not discourage him from taking on the challenge.

"One of my father's examples was Tamás Széchy," Susan Polgar told us when we sat down with her in her office at SPICE. "He was the coach of a whole generation of Hungarian champion Olympic swimmers." Széchy, sometimes known as 'the swimming Pope', coached his athletes to great victories: 15 Olympic medals (including 8 gold), 21 World Championship medals, and 30 European Championship medals.[205] "And believe it or not," she added with a laugh, "the guy didn't know how to swim. If he had fallen into a swimming pool, he would have drowned."

If not incredible swimming skills, what does a star swimming coach need? "It's the methodology," Susan explained. "It's the discipline. It's the approach. And that's transferable, in fact, to any field. The elements are pretty much the same."

Still, with his tiny pupil starting from scratch, Laszlo knew he would need to improve his understanding of the game. Prior to Susan's interest, he had never read a single book on chess. Now he went about reading, collecting and cataloging thousands of books on every aspect of chess and chess strategy. His own intense preparation, research and dogged determination eventually led him to write the iconic book, *Chess: 5334 Problems, Combinations, and Games and Reform Chess*.[206] Today, it is considered a classic for its wide survey of chess variants.

After six months of intense training, Polgar decided it was time to evaluate Susan's progress. He took his daughter to a well-worn Budapest men's chess hangout. Into a thick cloud of cigar and cigarette smoke—not exactly the haunt of most preschoolers—Polgar proceeded, holding Susan's hand, with a pillow tucked under his other arm. He wasted no time in challenging one of the locals to a chess match with his four-and-a-half year old daughter.

Finally, one man half-heartedly allowed Susan to sit for a game. Polgar placed the pillow down. Imagine what the seasoned old chess players must have thought, watching young Susan take her seat on the pillow—too small to reach the table without it. Imagine how they felt when she went on to systematically pick the veteran player's strategy apart, beating him handily. Then she did the same with several of his compatriots. Clearly, something unusual was going on.

Within a year of Susan's triumph at the chess club, her sister Sofia was born. Two years after that came Judit, the final Polgar sister. The three girls lived and breathed chess. They were homeschooled in all the required subjects, but Susan recalls an extensive daily chess-training regimen.

All of that work paid off. In 1992, at the age of 21, Susan Polgar became the first female ever to be named a Grandmaster by the World Chess Federation. That same year, her sister Judit became the youngest Grandmaster in history, overtaking Bobby Fischer's prior claim to the title by a month. Judit was fifteen at the time.

In 1972, Grandmaster Boris Spassky, considered at the time by many to be the best player in the world, was famously undone by the quirky American phenom Bobby Fischer. But in 1993, at the age of 56, Spassky also lost a $110,000 money match to another young Grandmaster: 16-year-old Judit Polgar.

Covering the match for the *Chicago Tribune*, writer Linnet Myers quoted an exhausted Spassky toward the end of his long two-week march to defeat. "I got the impression that I was a punched sack," he said, "and Judit was just beating me." After the defeat, Spassky was asked what his plans were. He said, "I want to sleep and sleep and sleep, like a winter bear. I'll dream that I'm 16 years old, and Judit Polgar is 56." After the match, Judit, seemingly nonchalant about defeating the revered Grandmaster said, "It was a pleasure to play such a great man, he's probably one of the best players of all time."[207]

By 2000, all three sisters found themselves ranked among the top ten female players in the world. Judit was ranked first, Susan was ranked second, and Sofia, sometimes jokingly described by the press as the "weak link", was sixth.

"Many coaches observing us when we were children believed that Sofia was the most talented, when it comes to pure chess knowledge," Susan told us. "I do believe that. She is very talented, but perhaps not as diligent as Judit and I were. And in fact, about Judit in particular, even though as a player, she is the most successful female player of all time, as a six-year-old, she seemed like the least talented."

How did the lagging youngest child go on to become a Grandmaster just nine years later? "She had a passion for the game," said Susan, "and passion can build amazing things."

Judit would go on to defeat reigning Grandmaster champion Garry Kasparov in an exhibition match in 2002. Just before the match, Kasparov unknowingly maximized the irony by publicly asserting that women just didn't make good chess players. At the time Judit defeated Kasparov, she was ranked number 8 overall in the world. She would continue to hold her number 1 woman's ranking right up to her retirement in 2014 at the age of 38.

So how did Laszlo and Karla manage to cultivate in their three daughters what some would call genius level chess play? Was it luck, an outstanding support network, environmental factors, superior strategic knowledge, positive coaching, tough competition, or deliberate practice?

And the answer is an unequivocal "yes". It's likely it was all those things because what some might call genius can be precipitated by a variety of factors. And those factors can have significantly different impact depending on the individual.

Isolating influences becomes enormously difficult when it comes to the complex behaviors of skill acquisition. Even the best scientists struggle with confirmation bias, the tendency to tilt data toward the expected conclusion. We are not robots operating off a simple set of universally programmed instructions; it's why every one of us is truly unique. At the same time, research does seem to suggest we can increase our chances of success by employing certain techniques and principles.

There are no guarantees, however, and as we've discussed in Chapter One, luck is one of those variables that you can't control. Were the three sisters lucky to be born to Laszlo and Klara, to be born with a plan mapped out in detail for their futures?

From the outside, you may find yourself wondering about the ethics of experimenting on one's own children. The phrase "human guinea pigs" comes to mind. Certainly the Polgars set their daughters on a lifelong trajectory that brought them into the public spotlight and took them far from the experiences of their peers.

However, when we talked to Susan, she said with confidence that she considers herself very fortunate, and that her sisters do as well. Serena Allott of the *Telegraph* said after interviewing Judit Polgar, "She accepted her childhood completely, as children do."

The Polgar sisters have had extraordinary lives. In a time and a place where travel was rare, tournaments brought them all over the world. In that same *Telegraph* interview, Judit remembered that, although at eight, she didn't play with toys, "I could say that I had been to Australia and held a koala and stroked a kangaroo."[208] Back home in their native Hungary, they were hailed as heroes.

And it may be worth noting that, although all three sisters have retired from the competitive circuit, they have each chosen to stay involved in chess. Judit and Sofia develop chess-based educational material for children in Hungary. Susan, along with husband and business partner Paul, put in long hours at SPICE.

As a chess coach, Susan devotes much of her career to replicating—and expanding on—the training regimen she experienced as a child. Even without the advantage of starting with impressionable four-year-olds, she has seen considerable success. Under her watch, the Webster University team became the first-ever chess team to be ranked #1 in the nation the same year it was founded. It has stayed at #1 ever since.[209]

Susan readily acknowledges that, when it comes to coaching, she had an excellent mentor in her father. When Laszlo Polgar embarked on his genius experiment, he was sailing into largely unchartered waters. The idea of building genius, as opposed to discovering inborn genius, was itself a novel concept. With the exception of his *Bring Up Genius*[210] treatise, which was untested up till that point, there was no manual to follow, no established roadmap or system to be leveraged. In hindsight, it's clear that many of Polgar's strategic insights share a commonality with the best teachings gleaned over the years from the greatest coaches across a wide variety of domains.

In fact, the story of the Polgar sisters is a fascinating case study of many of the factors that can go into the making of a genius.

Support Network

Since the Hungarian government took a dim view of homeschooling, Laszlo was constantly fending off their requests to put his daughters in the public school system. At one point, an armed policeman showed up with the intention of hauling Susan off to school.

Laszlo and Klara eventually agreed to have the girls tested once a year to demonstrate they weren't falling behind in their general studies. They passed with flying colors, showing particularly high-level skills in advanced math, and proficiency in several languages. It turns out both subjects were an essential part of Laszlo and Klara's home curriculum.

Although it's clear that he compromised with the Hungarian school system, Laszlo stuck to his guns about his very specific curriculum and methodology. The family's constant pushback against the authorities undoubtedly created an "us against them" mentality, driving the family closer together and, as a result, reinforcing their support network. Many scientists believe that feeling like part of a close team is a crucial factor for success.[211]

Moreover, the Polgar sisters seem to have built their own smaller support network amongst each other. A lone girl growing up immersed in chess could easily feel isolated from her peers—as well as from the male-dominated chess community. However, well before the Internet made it easy to find like-minded hobby enthusiasts, each Polgar sister always knew at least two other girls uniquely able to understand her highly unusual, specific experiences.

Mindset

The Polgar sisters were undoubtedly champions, but Susan remembers an approach that rewarded effort far more than it rewarded the trophies piling up. When they did lose, they were raised to not feel shame, but rather to use the experience as a means to understand and analyze opportunities for improvement. Consistent with Growth mindset, the loss provided the fuel and impetus for identifying areas of their game that needed strengthening.

"The dream was, become a world champion, become a Grandmaster," Susan said. "The long-term goal. But I think it's really important to enjoy the process, and believe in the process. And that, I think, was one of the most important things I got from my parents—that belief that hard work will result in fruits."

This fundamental belief—that training and dedication would ultimately prove more important than effortless, natural wins—precisely aligns the Polgar parents with Carol Dweck's growth mindset.

Access

The Polgars' living room, their inner sanctum sanctorum, was described as a shrine to chess, with chess paintings on the walls and chess boards in every corner of the room, along with a world class chess library of as many as 5000 books.

They had a chess catalog of over 200,000 games, which, in a time before home computers, was sorted by an enormous index card system filling an entire wall. It chronicled not just Grandmasters' past games, but also competitors' mistakes, opening moves, and strategies. This allowed the sisters to systematically deconstruct every aspect of the most famous chess matches every played—a key aspect of deliberate practice, and an important step toward building a more finely grained mental representation.

This availability of information, as well as the constant sensory reinforcement, kept the game at the forefront of their minds. Scientists would describe it as biasing the brain toward the importance of chess. As Judit Polgar said, "It just seemed natural that I should play chess, the apartment was full of books about chess and people playing chess."[212]

Many have suggested that the Polgar sisters are outliers, that there's a reason most chess champions are men, but to Susan, it mostly comes down to an issue of access. By and large, she says, young girls are not exposed to chess. Lack of early opportunities and support gives them a lasting disadvantage in competitive play.

Simply put: before you can begin your analysis and deconstruction—before you can even decide on a plan—you need a certain level of access to the necessary tools and information.

Variety in Practice

In a typical day, the sisters were tutored in a bevy of subjects with a special emphasis on languages and university level math. They also devoted up to 6 hours a day, 7 days a week, to sharpening all manner of chess skills.

It was an unusually efficient use of time for a trio of children. As Susan once said, "My dad believed in optimizing early childhood instead of wasting time playing outside or watching TV."

However, that is not to say that their practice time was an undifferentiated grind. The Polgar sisters engaged in a variety of different activities. They played against each other, they trained with chess tutors, they solved puzzles to improve their tactical and calculation skills, they played against elderly male Grandmasters, they played blindfolded speed chess, they worked on single moves and studied Grandmasters' strategies, they memorized masters' games, they studied many endgames, and they played in club tournaments.

Basically, they were taught to attack the subject from many different angles at once, always searching for creative new ways to approach the game.

Time Out

It wasn't all work and no play for the Polgar sisters. The morning would start with two or three hours of table tennis. They would sometimes take twenty-minute breaks solely devoted to telling jokes. Occasionally they swam and took side trips to theaters and museums. They also, of course, took much longer trips, traveling the world for tournaments.

Crucially, Laszlo recognized the need to keep his daughters moving forward without falling victim to burnout, a very common side effect when passion and practice go unregulated—perhaps especially in children who have been labeled prodigies.

Tennis champion Andre Agassi's famed burnout and mercurial relationship with his father/coach might serve as a cautionary tale of what happens when too much emphasis is placed on practice and winning.

Agassi accused his father of ruining his childhood by driving him too hard toward a tennis career, at the expense of nearly everything else. Agassi's father reportedly said he regretted nothing about his driving ambition, except that he would have in retrospect, pushed Andre toward a career in professional baseball or golf where salary opportunities are greater.[213]

Extreme Focus

Whether playing or coaching, Susan Polgar stresses the importance of concentration. "You cannot be impatient," she told us. "And that's the thing, other than practicing. That psychological element, to be in the right frame of mind." And not just being in the right frame of mind, but staying there. She pointed out that, while a footrace can be over in two minutes, a single chess game can last more than six hours. It is a true feat of endurance.

Continuing to sustain your A-game for that long is its own challenge, a definite test of rage to master. This is part of the reason that, perhaps surprisingly to some, Susan's current program recommends that young chess players also develop an exercise regimen. "We very strongly encourage it to our students, especially the top ones, because it develops the willpower and the endurance, and the discipline."

"And it's not easy, even physically, to sit for four, six hours," she added. "Sometimes, in collegiate tournaments, you play two games a day. Those games can easily go ten to twelve hours, and they do. So that physical fitness helps in chess as well." Strengthening the body allows a chess student to better shut out the distractions of an uncomfortable chair. "It's the extra," Susan explained, "between almost-winning and winning."

Competitive Strategy

Judit was described as a fierce, take-no-prisoners player. Her strategy was built on exploiting her opponent's weakness. Through pre-match preparation and keen in-game observation, she delighted in ferreting out her opponent's Achilles heel and using it against them. This meant her moves tended to be less predictable and often a little unconventional.

Her strategy very much echoes tactic six in the military treatise *The Art of War*,[214] written around the 4th or 5th century and sometimes attributed to famed Chinese strategist and general Sun Tzu: "Do not repeat the tactics which have gained you one victory, but let your methods be regulated by the infinite variety of circumstances."

This constant readjusting of strategy is part of what Susan loves about the game. "After a certain amount of training, anybody can be told to solve a specific problem," she said. "What's not so easy, and takes time and maturity, is how to apply that knowledge to something abstract. The game starts from the same position, but by move fifteen, twenty, it's virgin territory. It has never happened before. You have guidelines to guide you, but each game is going to be a new circumstance. Just like in our lives, right?"

Rage to Master

The intense desire to conquer a particular domain can serve as motivation to prepare and practice. This is true for a wide variety of people, across many different fields. When it comes to chess, Susan and Judit Polgar are prime examples.

Writer Carlin Flora reports that Ellen Winner, a psychologist at Boston College says:

> "The rage to master is a prodigy's primary motivation. Mastering a certain activity is more important to them than socializing, than anything else. You can force your kids to work harder, but can't get them to have the level of passion. The sisters could have just as easily rebelled against Laszlo."[215]

Another factor helped spur the sisters forward. Especially in the early days, the Polgars encountered considerable sexism in the chess community. They were sometimes ridiculed or dismissed, told to go home and play with dolls. From time to time, a prominent male chess player would publicly assert that women lacked some inherent quality necessary for true mastery. "It's not necessarily that the guys are chauvinistic pigs," Susan explained. "It's their real belief."

At 12 or 13, she had a conversation with one of the top Hungarian female players about whether a woman could one day become a Grandmaster, which has stuck with her ever since. "She was a friend of mine, but she was simply saying, 'I really don't believe it's possible.' I guess, in that generation, back thirty or forty years ago, women heard it so much, that we are not hardwired that way, that women actually believed it. It was stunning for me, to hear it from a woman, and to realize: people really believed it."

When we asked if that made her question herself, her response was immediate. "No. No. I was just stunned. It actually motivated me to prove them wrong."

It might be said that earlier generations of female chess players had been raised with pre-Bannister Effect thinking: they received the message that, since no woman had ever risen to the rank, it must be impossible. Susan, like Bannister before her, broke through that wall, paving the way for others. In 1991, she became the first woman to earn the Grandmaster title on the same basis as men. (Georgian chess champion Nona Gaprindashvili had been awarded the title in 1978, but only due to a special judgment from the World Chess Federation.)[216]

Susan's younger sister, Judit, followed in her footsteps just nine months later, to become what was at the time the youngest Grandmaster in history. How had Judit, at six considered the least talented, come so far? Perhaps Judit simply had developed the greatest rage to master, but also benefited from her older sister Susan's achievement via the Bannister Effect.

As we saw in Chapter One, birth order can be a competitive factor; often the youngest feels the need to prove their worthiness. The 'underdog effect' seems to have taken hold of the youngest sister Judit, with historic results.

Chunking

Chunking is a very efficient way to hack your short-term working memory system by creating specific memory combinations or patterns. For example, the number 09112001 is just beyond the reach of most people's ability to immediately memorize when viewed as a single string of digits. If you break it into chunks, however, 09-11-2001 instantly becomes far easier to remember: September 11th, a particularly meaningful and sad day in American history. Proper chunking creates context; and the associative nature of the brain means it is far more efficient when it works from a point of context.

When the Polgar sisters were memorizing sequences of moves from chess games past, they didn't need to commit to memory the discrete location of each single chess piece. Rather, they learned how to break the board down into smaller patterns, recurring configurations of moves. In other words, they were memorizing context clues.

Thus, neural code built from past matches becomes chess chunks; and opening with the right sequence of chunks allows one to duplicate Boris Spassky or Bobby Fisher's greatest hits, or to learn from their past defeats. The key is remembering the chunks and sequencing them properly.

The Polgar sisters can glimpse an in-progress chess game for a few seconds and perfectly recreate it from memory—a skill that also allows them to win games blindfolded, as long as they hear their opponent's moves spoken out loud. However impressive this may be, it is not a superpower, just high fluency in a language composed of aggregated chess moves. Show them a chessboard with pieces in an entirely random configuration, and their recall is no better than the most rank chess amateur.

This is true in much the same way that a haphazard string of letters will be much harder to remember than a short sequence of actual words. For example, "Lliks, sdliub noitanileym dna noitanileym, sdliub noititeper," reads like gibberish, and would be pretty difficult to memorize. But restructure the order and placement of the letters in each word and you get "Repetition builds myelination and myelination builds skill."

Context is critical in order for the brain to store, process and use information. Taking advantage of the brain's ability to create a database through chunking pays off, for everything from learning a language to excelling at chess.

Neuroscientist Daniel Bohr, in *The Ravenous Brain,* explains the power of chunking.

> "Consciousness and chunking allow us to turn the dull sludge of independent episodes in our lives into a shimmering, dense web, interlinked by all the myriad patterns we spot. It becomes a

positive feedback loop, making the detection of new connections even easier, and creates a domain ripe for understanding how things actually work, of reaching that supremely powerful realm of discerning the mechanism of things. At the same time, our memory system becomes far more efficient, effective — and intelligent — than it could ever be without such refined methods to extract useful structure from raw data."[217]

Neural Hijacking

In 2007, Professor Joy Hirsch of Columbia University, a specialist in brain imaging techniques, wanted to understand how Susan Polgar had radically expanded her own brain's chess database.[218]

In the back of your skull, you have a piece of neural matter called your fusiform gyrus, often referred to as your fusiform face area. Until fairly recently, it was thought that the fusiform was almost exclusively designed to recognize faces.

Many species, especially humans, rely on the ability to recognize and read faces in everyday encounters. Roughly a third of the human brain is dedicated to processing visual information.[219] If your fusiform face area gets damaged, you walk around unable to recognize anyone.

People with prosopagnosia, sometimes called face blindness, are literally unable to pick their mother or best friend out of a lineup. It's estimated that 2% of the population has some level of face blindness.[220] On the other hand, Scotland Yard in London employs a handful of "super-recognizers," individuals with extraordinary face recognition abilities, for scouring police photos and crime videos to help identify criminals.[221]

For the average person, if everything is working properly, the fusiform gyrus allows for storage of up to 10,000 faces in long-term memory. So what does the brain's facial recognition software have to do with chess expert Susan Polgar? It turns out that the fusiform face area is not just a cool face recognizer, but a sophisticated pattern recognizer, says scientists Michael J. Tarr and Isabel Gauthier.[222]

When neuroscientists initially examined the fusiform face areas of most experimental subjects, specifically those who weren't experts in any particular field, they hypothesized that the fusiform face area applied only in faces. Face recognition is the only kind of deep domain knowledge that we all seem to store and access regularly. Yet when scientists put Susan Polgar in an fMRI and had her play chess in real time, activity in her fusiform face area suggested it was being used as her mental chess library.

An expert like Susan Polgar has built, and relies on, a huge memory bank of chunked chess patterns. By the age of 10, most children have a vocabulary of about 10,000 words. It's estimated that Susan at the same age had already built a chess vocabulary of approximately 100,000 chunks.[223] The fusiform face area was perfectly suited to store these chunks for later access.

Most of us never need that kind of additional warehouse space because we don't build the massive inventory of specific memories that a chess Grandmaster like Susan needs.

In any given chess game, within four moves alone, the possible move options number at an astounding 288 billion.[224] It would be impossible for Susan, or any human, to operate without special brain pattern recognition software—or more accurately, given the electrical/chemical nature of the neuron signals, wetware.

People who have developed deep expertise in a given domain, from birdwatchers to car experts, can take advantage of the fusiform face area's ability to identify patterns.

Think about an NFL quarterback dropping back in the pocket and instantly reading the defensive play pattern unfolding around him as three-hundred-plus pound linemen bear down on him. Consider trumpet player Dave Guy's ability to hear a few notes, recognize the musical key, and instantly improvise a harmony pattern without skipping a single beat.

Beyond Chunking

However, Anders Ericsson cautions us not to overstate the importance of chunking. He points out that simply memorizing chess chunks does not guarantee any improvement in your playing skills. The ability to predict the effects of a long string of moves and countermoves cannot be explained by simple pattern recall. You must also train yourself to understand and evaluate the data —to build a detailed and responsive mental model with cues that will let you efficiently call up essential information during task performance.[225]

In fact, highly skilled experts don't always have exceptional working memories. For example, studies show that experienced doctors recall less overall data about their patients than novices, but expert doctors are more accurate in their diagnoses. In other words, it's not the quantity of information stored, but the ability to recognize what is important to the patient's health.

Ericsson suggests there are a couple of keys to raising your skill level:

First, pinpoint the skills or behaviors that are essential to expert-level performance. For example, perhaps the best metric to judge chess expertise is the speed and quality of a player's move decisions. The essential question becomes, "What are the experts doing differently than me?"

Second, identify the mechanisms that can help you learn how to perform better. For instance, in chess, studying Grandmasters' past games allows you to "play along" with a match, trying to anticipate the best moves and comparing your choices with those of the masters. In this way, you can begin to learn, copy, and eventually intuit qualitatively better decisions in your own playing.

This is why coaches are so important in just about every skill endeavor. By working with a coach and reviewing resources like past chess games, game films, and recorded musical performances, you can piggyback on the skills of masters. As the old saying goes, 'there is no need to reinvent the wheel.' Study of the experts helps you improve your own mental representation for better performance.

So it should be no surprise that studying Grandmaster games correlates better to improved skill level than just playing friendly matches with one's pals.

True mastery comes not just from recognizing and memorizing information, but developing the ability to sort, interpret and manipulate the essential data in real time. These skills are made manifest through deliberate practice. In Ericsson's view, expertise is more than storing chunks of key information in memory—it's about incorporating the deep domain knowledge you can practically apply when confronted with a task, problem or situation in your field.

Anders Ericsson says that the domains that lend themselves best to deliberate practice are the ones with years of documented training methodologies and objective ways to assess improvement. These include domains like classical music, sports, math, and chess. Domains like jazz, painting, or business, where excellence is more a matter of opinion, make it tougher to leverage each and every one of the deliberate practice techniques.

Still, employing any of the techniques discussed gives you a leg up. Taking the time to build a deliberate practice app not only keeps you out of stagnation swamp; it's required in order to myelinate for skill acquisition at the highest level.

In some ways, this might be that seemingly superhuman quality we see in all of our heroes: their willingness to dedicate time and effort in pursuit of their long-term goals. Maybe more than anything, we admire the determination and focus that drives their personal kaizen event of incremental improvement toward unattainable perfection. This is what sets the so-called geniuses apart.

Not all that long ago, the path to deliberate practice was largely the secret dominion of the genius club. It's not a secret anymore.

Chapter Key Points: Hacking Your Deliberate Practice App

- To create quality neuron connections for any behavior, you'll need time and error-free repetition.

- Expertise is built through deliberate practice

- To engage in deliberate practice:

- Deconstruct the process and identify all the pieces and parts that make up the whole.

- Make sure you have an accurate and detailed mental representation of your goal.

- Pursue perfection. When you encounter any problem, ask yourself what is causing the issue.

- Have a domain expert (coach) verify your plan. With the help of your coach, craft a deliberate practice solution to address each piece or part separately.

- Practice in concentrated, highly focused 15-minute sessions.

  ◦ Make sure you've gained absolute qualitative mastery over a particular piece or part before moving on to the next on section.

  ◦ Practice the final production (the aggregation of all the pieces and parts) under real life conditions until it's second nature.

- Have a domain expert regularly review, tweak, and help you vary your practice regimen.

Chapter 5
Habit Formation App

So far we've stoked the fires in the *pit of practice* and fanned the flames of our Rage to Master. Now it's time to convert our deliberate practice regimen into a full-fledged brain app, or habit. (We'll be using those terms interchangeably from here on out.)

Apps are what make your smart phone smart, and in much the same way, brain apps are what allow you to take better advantage of your brain's potential. Practice is crucial in creating new habits.

By now, you understand the basic process: practice builds repetition, which myelinates connections, which in turn, form habit. Habit is really nothing more than myelinated neural code put into action. But, of course, knowing how something works is not the same as actually doing it.

Brain Apps in Action

An estimated 47% of your daily behavior is driven by habit, bits of neural code that your reflexive System 1 uses to save analytical System 2 from expending unnecessary energy.[226] From flipping on a light switch to driving a car, you've built a whole series of brain apps that you rely on to function. And those simple bits of neural code can be stacked on top of each other to create more elaborate behaviors. The formation of those behaviors is difficult to detect because they generally run below your conscious awareness, but it's still happening all the time.

Psychologist B.F. Skinner famously illustrated this point when he taught a rat to react to the Star-Spangled Banner by hoisting a miniature American flag and saluting with its front leg.[227] The rat clearly did not attach meaning to these gestures; Skinner had taught the rat a series of actions, which, when combined, created the impression that the rat was actually hoisting and saluting with intention. The rat had simply learned, through repetition and reward, to respond to certain cues.

Once a behavior becomes automatic, much of what we do is in a sense rat-like; we don't actually *think* about it, we just perform on cue.

For instance, it's unlikely you set out to watch TV for the typical three to five hours each night[228] and then actively practiced and perfected the technique until you mastered sitting in front of a glowing screen. Still, through repetitive action, your reflexive System 1, in essence, was busy practicing and building that TV app. It works the same way for Internet browsing, too.

If you stack your Internet app on top of another app, say evening snacking, we can start to see how the power of unintended habit programming has a dramatic effect on both our watching and our waistlines.

Habit Stacking

Habit stacking is the combination of several habits that trigger each other or are triggered at the same time. Although excessive snacking may be undesirable and can have consequences to our over-all well-being, other examples can have a far greater impact on our lives.

Imagine what happens if you've built habit stacks out of a variety of fixed mindset positions. If you've habituated operating out of a position of fear, and at the same time, you have a habit of shutting out critical input from others, you can see how habit stacks can quite literally stack the odds of success against you. Then, when things don't turn out well, it feels as though there are unseen forces working to bring you down. This, of course, is correct; it's just that those forces are coming from within.

Studying habits work provides us with some important opportunities. First, we can trace behaviors back to the source and better evaluate their meaning. Second, we can consciously rewire neural code to offset the flaws and glitches in our current brain apps. It gives us a better chance of becoming, as Pulitzer Prize-winning author David Foster Wallace once put it, "the lords of our tiny skull-sized kingdoms."[229]

Many of our habit stacks have built up slowly over our entire lifetime. This is because our operating code is being written, overwritten and amended every hour that we're awake—and in a very real sense, during memory consolidation while we're sleeping. From simple to elaborate patterns, we live by internal rules that often drive predictable, habituated behavior.

Again, it's important to remember that if you recognize your own behavior, you can take a far more active role in consciously constructing your own code.

When Your Habit Stacks Are Turned Against You

There are plenty of great sci-fi stories about sinister people manipulating the brains of everyday folk for their own nefarious purposes. Of course, that's science fiction, but what if some other entity, evil or not, also had the ability to hack your brain's habit code? It's not as far-fetched as you might think.

It all starts with metadata, an enormous repository of collected information about, for instance, common buying behaviors. Retailers and marketers are constantly scooping up such metadata, looking at how you've bought things in the past to predict how you will make such decisions in the future. Here we have an excellent example of how marketers are leveraging your habit stacks for their own monetary gain.

As *New York Times* writer Charles Duhigg recounts, one particularly savvy statistician named Andrew Pole provided that kind of leverage for the chain store Target,[230] with somewhat eerie results.

Target, it seems, had figured out the value of the childbirth cycle. From disposable diapers to strollers to organic baby food, it turns out that from a retailer's point of view, gaining the brand loyalty of a new mom really is like hitting the mother lode. With Pole's number crunching, Target ran "predictive analytics." By tracking purchase history of items like unscented lotion and prenatal vitamins, which tend to follow certain distinct patterns, Target was able to make very good guesses about which shoppers were pregnant, down to the stage of pregnancy.

Armed with this intelligence, Target would, well— *target* the future mom. Where one might assume everyone on your block was getting the same sales flyer, in fact the company was sending some pregnant women specialized coupons and flyers with ads based on their current and future baby needs. This gave the mass merchandiser a leg up on the competition for the hearts and minds of expectant mothers. "We knew that if we could identify them in their second trimester, there's a good chance we could capture them for years," Pole told Duhigg.[231]

Duhigg notes that Target was also working another habit angle, based on the research of Alan Andreasen. In the 1980's, UCLA professor Alan Andreasen ran a study to examine people's purchasing habits regarding everyday products like toothpaste, soap and toilet paper.[232]

Andreasen's research showed that once a consumer's routine solidified into unconscious autopilot, shoppers were pretty much locked into buying a given brand. This, of course, was good news to stores like Target, provided they could establish that habit in the first place.

But there was another powerful piece of learning that came out of Andreasen's research. He found that when people underwent major transitions—a death in the family, moving, a new job, marriage, divorce, or, yes, having a baby—these adjustment periods made regular routines much more malleable. Changing someone's habit became more possible.

This was why Target was interested in catching consumers during key life events like pregnancy, when established buying patterns at non-Target retailers could be switched in their favor.

The company was on a roll until they unwittingly began to send coupons for baby-related products to a high school girl. Outraged, her father cornered the manager of his local Target store. But several days later, the father made an apologetic phone call to the manager after his daughter confessed that she was indeed pregnant. By chronicling her purchases, Target had awkwardly shoehorned itself into a momentous event in one shopper's life—and in so doing, demonstrated another truth in business: the law of unintended consequences.

Of course, adjusting a group's habits can be beneficial in a variety of ways. When former U.S. Treasury Secretary Paul O'Neil was CEO of Alcoa, he helped the struggling company return to prominence on the Dow Jones Exchange by improving safety-related workplace routines. It led to a cultural shift in the organization and a renewed financial vitality. One of the factors that helped Coach Tony Dungy take the formerly terrible NFL Colts to the Super Bowl included focusing on habits and carefully altering his team's reaction to "on-field cues."[233]

And according to Duhigg, it's not just mass merchandisers, companies and professional sports teams that are leveraging habit. President Obama had a 'habit specialist' on his campaign team to help analyze voting patterns and trigger new ones.

Habit Time

You might have heard that it takes 21 days to build a habit.

Psychologist Jeremy Dean, author of *Making Habits, Breaking Habits*, says that there is only one problem with 21 days as the habit gold standard.[234] It's less of a statistic and more of an urban myth.

Dean believes the myth originated with plastic surgeon Maxwell Maltz, M.D. who in 1960 wrote the best seller *Psycho-Cybernetics*.[235] The book sold 30 million copies and generated a lot of memorable quotes. One of those oft-repeated quotes is that habits take 21 days to stick. This was supposedly based on Maltz's observations from his medical practice that amputees and facelift patients both took about 21 days to habituate to their new identities.

In his book's preface, Maltz elaborates on this idea:

> "People must live in a new house for about three weeks before it begins to 'seem like home'. These, and many other commonly observed phenomena tend to show that it requires a minimum

of about 21 days for an old mental image to dissolve and a new one to jell."[236]

The bottom line, however, is that Maltz's observations were not based on any formal scientific study.

For a more scientific approach to understanding what a habit timeline looks like, we turn to University College London. Researchers ran a habit acquisition study with 96 people over 84 days.[237] The participants logged into a website when they had reached a level of 'automaticity,' meaning their behavior had become routine. The amount of time it took to acquire a habit fluctuated quite a bit based on the difficulty of the behavior they were trying to master.

Drinking a glass of water after breakfast took only about 20 days to habituate, surprisingly close to Maltz's observation. But the average habit, across all study participants, took about 66 days to lock down. More difficult and challenging habits, based on extrapolation (since the study only ran for 84 days) worked out to be closer to 254 days or more.

Understanding Habit

Three features define a habit. First, when we are running habit code, our reflexive System 1 has taken over and we are operating in autopilot, like when we're driving a car or typing. For the most part, we initiate a habit without deliberation and the habit generally operates below our awareness.

Second, because we are operating out of System 1 automation mode, emotion researcher Nico Frijda says that habitual behavior tends to be largely devoid of emotion.[238]

No experience is as powerful as the first time you encounter it. Most habituated experiences, performed again and again over a period of time, diminish in impact and lose some of their poignancy. So even things like sex, drugs, rock and roll, or the stress associated with hearing police sirens can lose emotional gravitas.

The tamping down of excitement allows the brain to burn less glucose. Thus we could say on the plus side, habits are more energy efficient. On the negative side, the humdrum repetition of habits means they can become stale and boring. And as we've seen with the Target example, under the right conditions, habits are susceptible to change.

Lastly, we tend to initiate habits under the same set of reoccurring circumstances, much like Russian researcher Pavlov proved in his experiment with dogs.[239] When conditioned to hear a bell before feeding, Pavlov's dogs began to salivate in anticipation of the food. Humans operate in a similar fashion; our habits are strongly rooted in, and triggered by environmental context.

Unintentional Programming

From the time of your birth, your parents were bombarding you with repetitive messaging, everything from "Eat your peas" to "Do your homework." Messaging is the very essence of being a parent. And that external messaging is a partial key to whom you grow up to be. But what about the internal messaging: how does that happen?

The mass of 200 million interwoven fibers linking your brain's left and right hemisphere is known as your corpus callosum.[240] This high-speed communication bridge ensures the two hemispheres work in sync with each other, neurons connecting parts that handle vision, hearing, spacial reasoning, and thought. As you experience the world, your brain is constantly learning and adjusting in an effort to maximize fuel efficiency and keep you alive.

Interestingly, these neural adjustments happen behind the curtain; for the most part, we are unaware of them. Rewiring tends to follow patterns. As we've discussed, the basal ganglia converts repetitive patterns into little bits of myelinated neural code to ensure that a particular brain application lives on in your unconscious.[241] This frees up your working memory from trying to remember to execute a particular task like flossing.

However, your basal ganglia doesn't have a horse in the race as far as your mental or physical health is concerned. It takes no side in whether the habituation it programs is good or bad. Balancing your checkbook every weekend, practicing the piano five hours a week, hitting the bars every Saturday night —all feels the same to your basal ganglia. It turns autopilot on whenever you give it enough repetition.

Of course, the same is true if you start walking after dinner in the evening. The brain makes little distinction between healthy endeavors and the unhealthy ones. When it finds a pattern, it's like a bear finding a picnic basket: the feast is on.

What's interesting is that so much of what we habituate is unintentional. For example, there is a pretty high likelihood that tonight when you sit down for dinner (provided that you eat at home), you will sit in the same chair you sat in the night before. Of course, there's no intrinsic value to that specific chair. It's likely there are at least a couple more chairs nearby that look exactly the same.

But over time, your habit-driven brain has programmed that shortcut to free you from dithering each evening about chair choice. And now, brain app firmly installed, if someone plops down in your chair of choice, you might remind them rather indignantly, "Hey, that's my chair!"

Given our ability to create neural programming, you'd think we'd take huge advantage of our brain's flexibility and program the thing to the hilt. Unfortunately, there is one big hurdle necessary to hacking your own brain's habit formation.

Neural code, as we've learned, often takes at least two months to build. In today's fast-paced world, two months of repeatedly practicing something feels like an eternity. More advanced brain apps, like learning Chinese, can take many months of repetition. Intentionally building consequential brain apps is generally hard work. We tend to be more comfortable running the same apps, the ones we've already programmed.

For this reason, many of us find ourselves doing the same routines over and over again. We repeatedly eat the same foods, wear the same clothes, hang out with the same friends and do pretty much the same activities with those friends. It takes real physical and mental effort to radically depart from our daily routines. Take something as simple as brushing your teeth. Try brushing with your opposite hand—not only does it feel strange, but you'll find yourself quickly reverting back to the way you normally brush. Even simple routines are powerful reminders of the brain's reliance on existing habit code.

Wiring Brain Apps

Behavior Scientist BJ Fogg has developed a useful approach to changing your routine. He told us that when it comes to building habits, the key is to start small: very small.

The idea is to knock down the barrier to entry by making those first steps toward building a new habit so easy it's almost impossible not to do them.

Here's an example BJ Fogg uses: you probably brush your teeth every day. (No doubt, with the same hand.) But chances are pretty good you don't floss. Only about 40% of Americans regularly floss, according to the American Dental Association.[242]

Still, presumably everyone wants to avoid gum disease, and although the data is inconclusive,[243] some health professionals think there's a link between gum disease and heart disease.[244] They believe the mouth is one of the easiest ways bacteria can be invade your blood supply. If you get an infection, if you get gingivitis, that bacteria can find its way into your bloodstream—which is bad news for your whole system.

The theory says that flossing is one of the surest strategies known to reduce bacteria's access to your bloodstream. It shores up the gum line, creating a seal around your teeth that keeps the bacteria out. If you're worried about heart disease or the health of your blood supply, flossing is one app you definitely want to build.

Now, suppose you recognize the value of flossing but seldom do it. Your basal ganglia has already built a habit for brushing your teeth, so it makes sense to use brushing as your trigger, or anchor mechanism.

Start by putting your floss right beside your toothbrush to remind yourself what comes next.

After you brush your teeth, floss. But here's the catch: start out flossing only one tooth. Make the barrier to entry so comically small that you can't begin to psyche yourself out about it and pass the buck to Future-Self. After you've reached your first day's goal of just one flossed tooth, take credit for it; let yourself feel good about the progress you've made. Literally praise yourself out loud. As amusing as this might sound, by so doing your brain will be giving you a little shot of dopamine, the feel-good drug that helps reinforce new brain apps.

The next day, follow the same routine, again flossing a single tooth. It's important to give yourself time to make the act of flossing a habit. If you raise the bar to a mouthful of teeth at the beginning, you'll weaken the habit acquisition program and jeopardize the entire flossing enterprise. On the other hand, once you solidly establish that single tooth habit, you'll find yourself gradually adding more teeth into the program. Build up your regimen tooth by tooth, and eventually you'll be flossing all your pearly whites. Following this simple process, motivating yourself through verbal reward, allows you to cultivate what Fogg calls the "Tiny Habit Method." After about 66 days, the tiny habit will grow into a fully programmed, automatic brain app.

It's easy to get excited about a new goal and expect positive improvements to happen almost immediately. When we are impatient and don't appreciate the value in delayed gratification, we get discouraged. If morale flags, we're more likely to rely on Future-Self for making a potential change. Start small, and you can get Current-Self in on the action.

Beyond the smallness of the steps, you also make things easier on yourself if you can trigger the new behavior with something you've already habituated—it creates one more layer in a habit stack. Otherwise, you might legitimately intend to floss, only to forget about it until you're doing something else, like driving down the highway, making flossing, like texting, both difficult and inadvisable.

Tiny Hacks

When we spoke to B.J. Fogg, he offered the following advice:

1. Take advantage of neural programming you've already created by piggybacking on an existing habit. Your existing habit becomes the trigger for your new desired action.

2. Keep the motivation rolling by letting yourself feel good after you've successfully completed the desired action.

3. Start small—really small, like with the flossing example.

Fogg also said that the beauty of the Tiny Habits Method is how easily habits can grow into something more. "If you have a habit of putting on your walking shoes, eventually you'll have a habit of walking thirty minutes a day. The metaphor I use a lot is to think of it as a small plant, a little seed that you're planting... a little seed will grow to an expected size, as long as you keep it nurtured. And you can't force it to grow, without causing it harm. You keep it nurtured and you allow it to expand on its own."

Getting Started

Let's suppose that you're ready to build a new brain app. You've decided to start writing a weekly blog about your passion, travel. To reach that goal you're going to follow the **SMARTR** model from Chapter Two, which means there's some pre-preparation involved. First, you're going to have to hone your writing chops.

To that end, like Mark Twain before you, you're going to need to write an awful lot and get accustomed to hitting a daily word count. You understand that the growth will be exponential and it won't happen overnight. You're also aware that, again like Twain, you might not be an accurate judge of your own writing quality so you've joined a writing group to ensure a good feedback loop.

Being the clever person you are, you've ratcheted up your willpower ahead of time by making sure you're exercising, eating right and getting enough sleep.

In order to gain the necessary practice time for your travel blog, you've cancelled your TV cable subscription and moved your computer to the former TV area.

Why move your computer? You're taking advantage of habituated context clues. When you walk into that room after dinner as you normally do, now you'll piggyback on your old TV watching trigger to build your new writing app.

Fogg told us it's critical to work out the first step in your new routine.

> "What is the starter step in writing that blog? Is it opening up [the website platform] Wordpress? Is it taking out a notebook? And you would work to make that an automatic behavior."

> "For other types of habits," he continued, "you would scale it back to something very small. Not a starter step, but you would actually maybe write one sentence, and that would be the tiny habit. But I think, with regular blogging, the right thing is to bring it back to a starter step, not a tiny version. A tiny version would be writing one sentence, a starter step is opening up Wordpress."

Let's imagine that you've kicked off your habit by opening Wordpress. Understanding the power of feeling good in boosting your dopamine levels, you're already celebrating by dancing around the room. But you're not done yet. As a visual reminder, you've bought a big calendar on which you intend to put a red X on every day that you write for the next two months in order to both track your progress and as an incentive to keep the momentum chain going.

In fact, you're so serious about making the travel blog happen that you've gone one step further: you've enlisted a trusted friend to request regular updates on your progress, thereby making it much tougher for you to back out. Telling a friend was a pretty good idea, because research shows that with a support network you're much more likely to hit your goal.[245]

You've seemingly planned for everything except for one thing — the dreaded *extinction burst*.[246]

Extinction Bursts

You might be surprised to find that your analytical System 2 is up against an internal conspirator bent on thwarting your writing goal in favor of reinstating TV watching. The nemesis is your own primitive reptilian brain. This region, evolutionarily the precursor to your more advanced prefrontal cortex, has some habit wrecking shenanigans up its figurative sleeve.

The shenanigan in question is an extinction burst. Your reptilian brain cleverly waits to spring the trap only when you are specifically in the very final stage of habit change, your TV habit all but pushed aside by your new writing habit. An extinction burst is much like a Hail Mary play in football, in which desperation drives an all-or-nothing strategy for success. Your reptilian brain makes a final push to reestablish and save your TV habit from extinction.

It goes something like this: suddenly, you get a deep craving to watch an episode of *NCIS* instead of putting in your nightly word count. The X's on your calendar can attest to the fact that you've maintained your writing willpower up until that moment. But the urge to watch TV seems both ridiculous and undeniable. You begin to wonder how long it will take to get your cable service back on and you frantically search your smartphone for the number to call.

Feeling a level of weird desperation, you even find yourself contemplating inviting yourself over to your neighbors to poach a little TV. What's so strange about all this is that this TV desire seemed to drop into your thoughts full-blown and out of the blue, taunting you like the sirens in the Odyssey.[247]

You can thank your reptilian brain for ramping up the *NCIS* craving to almost unbearable level. This same extinction burst phenomena might explain why dieters succumb to binging behavior after they've been so diligent in their efforts to kick their sugar addiction.

Why would part of your brain thwart your intention?

Perhaps because your TV watching behavior has become so deeply ingrained and so hardwired through Hebbs rule, that your more primitive brain region misidentifies the habit as something vital to your very survival and sends out the signal to hold on at all costs.

Although watching television is, from a rational point of view, clearly not key to your continued existence, still your reflexive System 1's job is to follows orders; it's not there to evaluate the merits of the programming. Like the electric company during a power outage, the priority is to simply bring everybody's power back on line.

The theory goes that extinction bursts may operate similarly to certain allergies. Some allergies occur when your immune system misidentifies an allergen as an unwanted invader, ramping up histamines in the system to fight it off.

Since the brain's job can be viewed primarily as keeping you alive, the brain errs on the side of caution. Occasional misidentification is a small price to pay if overall the brain's survival strategy has proven effective, as it has in the past. However, extinction bursts are dangerous, largely because they hit without warning and can overwhelm all of your good intentions.

Odysseus solved his fear of succumbing to the tempting siren song by having his shipmates lash him to the mast of his ship. That might have worked for the legendary Greek hero, but ship masts aren't always easy to find, especially when it comes to the dessert aisle of your grocery store or the whispers of your favorite unwatched TV show. One strategy to avoid extinction bursts is to thoroughly invest in the habit process and understand how each mechanism is connected so that you can maximize your effort when wiring up your next brain app.

The Wiring Process

Behavioral change models are plentiful, but for our purposes, we'll rely on B.J. Fogg's system, the Fogg Behavior Model, or FBM.

Fogg writes, "The FBM asserts that for a person to perform a target behavior, he or she must (1) be sufficiently motivated, (2) have the ability to perform the behavior, and (3) be triggered to perform the behavior. These three factors must occur at the same moment, else the behavior will not happen."[248]

Suppose you want to kick your level of exercise up a notch. A trusted friend has told you about a great exercise facility close to your house. Included in its introductory offer is a customized strength and stamina assessment, along with a personal exercise plan. This program normally carries a $100 charge. To sweeten the deal, it's offering a 50% discount on a year membership, provided you sign up in the next five days.

You understand the power of loss aversion, that is to say, the brain bias we all carry that weighs the pain of a potential loss twice as strongly as the joy of a potential gain. In other words, skipping that deal feels worse than spending the money on membership.

Loss aversion is a potent force; it's what helped build Las Vegas into a world-renowned gambling mecca. Gamblers who are down thousands of dollars keep pushing for a win, because it feels so bad to walk away from the table with less money than they had at the start. We all know how this story ends.

And even though you understand the manipulative aspect behind the gym's offer, you still find yourself being pulled in by it. The promotion's expiration date becomes a powerful motivational trigger. In fact, when you stop by the gym for a preview, you notice they've reinforced this special "once-in-a lifetime offer" with a giant digital clock that's ticking down toward the end of the promotion time window like Time Square on New Year's eve. (Nice loss aversion touch, marketing team.)

The gym still has two hurdles to jump before they gain you as a member. The first is your motivation to sign up—your own perceived readiness to begin exercising. The second is your ability to pay.

Fogg points out that we can think about motivation and ability as each existing on a sliding scale. After the initial trigger, if both motivation and ability are high, there's a good chance you'll cross the decision threshold and sign up.

If neither motivation nor ability is high, it's unlikely you'll sign up. If motivation is high but your ability is low—for example, if you don't have a lot of spare cash and your credit is maxed out—you will likely skip the offer. It won't matter if the gym drops the membership rate another 25% as an enticement; all the motivation in the world can't overcome your empty bank account.

This is an important rule to understand: increased motivation alone isn't enough to change your behavior and wire for a new brain app. This is why motivational speakers can get an audience really fired up, and seemingly motivated, without spurring any lasting changes. If you lack the ability to follow through on the call to action, the money, and time you spent on the motivational seminar becomes a sunk cost.

However, if you clearly have the ability to follow through on the call to action, a tiny dash of motivation might be all you really need.

Imagine that you're at a carnival and you have a chance to win a new outdoor gas grill in a basketball-shooting contest. To win the grill, you need to sink five free throws in a row. Assuming you were a shooting ace back in high school, you might consider trying your hand at the contest, even though you don't really need a new gas grill. In this case, ability might trump motivation.

If we review the description of luck from Chapter One—being in the right place, at the right time, under the right circumstances—we begin to see a parallel. In the case of luck, the three aforementioned conditions happen naturally without directed human intervention and are therefore not subject to one's control.

When it comes to orchestrating change, triggers, motivation and ability do fall largely under your control. This is why the oft quoted saying "the harder I work, the luckier I get," is technically inaccurate, but illustrates a deeper truth. Hard work can't generate luck, but it can allow you to create a situation for yourself in which it is easier to succeed.

As we've seen, motivation and ability move on a sliding scale, but nothing can happen without a trigger, the cue or switch that fires up the necessary neural circuitry. Fogg says, "Successful triggers have three characteristics. First, we notice the trigger. Second, we associate the trigger with a target behavior. Third, the trigger happens when we are both motivated and able to perform the behavior."[249]

Triggers can come in all shapes and sizes. I might trust an alarm clock to get me up in the morning to go jogging. I might rely on the pangs in my stomach to remind me it's time to eat lunch. A tone on my smartphone might trigger me to check my email, or tooth brushing might be an existing trigger I can habit stack to build my new flossing app.

Not all triggers last. If I put a yellow Post-it note on the refrigerator to remind me to practice piano, it might be effective at first, but after a while, the note fades into the refrigerator scenery and I no longer give it import. This, Fogg says, is why a reminder is not the same as a trigger.

The best triggers require some kind of physical action and are linked with the actual task at hand. The buzz of an email alert is a good example of a trigger that is directly associated with a next step: checking your messages.

When wiring for brain apps, your goal is to create physical trigger associations. Putting an alarm clock on top of your running shoes, when your shoes are on the other side of your bedroom, means that when the morning alarm sounds, you're forced to get out of bed to turn off the alarm. At that point, the alarm serves as both timing trigger and cue that your running shoes are patiently waiting for you to hit the pavement. Setting up this scenario the night before makes it hard to escape the less-than-subtle double trigger effect.

Timing is key. Poorly timed triggers not only have little impact, but they can prove to be irritating. Fogg cites spam and pop up ads as relatively ineffectual Internet triggers. When they show up on the screen without warning, they seldom serve as a call to action, unless their call to action was designed to annoy you.

However, studying the Fogg model, it becomes increasingly clear how Amazon's 1-Click technology—the ability to order, pay for and download a book with a single click—has proven to be so effective. That button is an instant trigger for buying, aimed at people with the motivation to be browsing for that item in the first place, and as far as ability goes, what could be simpler than a single press of a finger? Companies like Amazon are leveraging the habit behavior into their selling scheme. Whether you think of it this way or not, with each successful click, they are actively complicit in triggering the rewiring of your brain.

Fogg breaks trigger mechanisms down into three types: sparks, facilitators and signals.

Spark Triggers

Spark triggers are highly emotional prompts that pull at your heartstrings to summon up feelings of hope, fear, guilt, or empathy. Their goal is to significantly ramp up your motivation.

Seeing a video of a starving child in a third-world nation is a spark designed to evoke empathy in order to maximize your motivation for charitable giving. A PBS pledge drive is a spark to trigger guilt if you continually watch their programming for free. (They usually couple this with some kind of sweetener or reward, like a CD set or book, to add a positive appeal as well.) Insurance companies often use images of disasters to trigger the fear of losing your belongings and increase your motivation for taking out a homeowner's policy.

Facilitator Triggers

Facilitator triggers are designed to take advantage of high motivation when the ability to carry out the call to action is in question. Facilitator triggers lower the ability hurdle by simplifying a process. Amazon's

1-Click technology is a facilitator trigger, making ordering a book almost effortless.

When your health insurance company mails you a bill that comes with a pre-labeled return envelope for simpler payment, they are hoping to simplify the process and get their money sooner.

Signal Triggers

Signal triggers aren't set up to motivate or lessen the ability required. Their goal is to signal that it's time to initiate a specific behavior. The ding on a microwave tells you to open the microwave door. An alarm clock serves as a common signal trigger. When motivation and ability are already high, adding a signal trigger is both necessary and an effective way to launch a new brain app.

With the advent of smartphones, triggers are taking on a far more significant role. The sudden immediacy of the Internet offers new opportunity to modify one's behavior. Before we were all carrying smartphones, much of the wiring or learning phase required physical access to certain resources—books, lectures, and domain experts. Today, we can easily watch a tutorial on a myriad of subjects by some of the greatest authorities in the world.

Smartphones also provide us with countless useful programs, allowing us to create and perfectly time our own triggers. The opportunity to take advantage of technology and wire your own bevy of brain apps has never been greater.

Simplicity Profile

Fogg writes that five elements often come into play when we consider the ability/simplicity factor of building a habit. He likens the five elements to links in a chain: each link must be intact in order for the chain to support the habit change and therefore, the new brain app.

Fogg's research shows your ability/simplicity chain is usually made up of the following links: time, money, physical effort, brain cycles, and non-routine. He calls this your simplicity profile.

Time

Building a brain app requires consistently using Chapter Two's time blocking process to set aside the necessary time for the task. On first blush, this might seem like simply making time to deliberately practice, but it also includes marshaling all of the resources needed, and/or setting up the circumstances to make the deliberate practice possible in the first place. This would include making sure a trigger is in place to initiate the process.

If my goal is to build a morning jogging brain app, I need to set the alarm ahead of time. As we discussed, I might consider placing the alarm on top of my running gear to double trigger my jogging activity. But I'll also want to already have all my running gear together and ready to go. This might mean I'll need to do a load of laundry the night before, and maybe clean the mud off my shoes from yesterday's morning run ahead of time.

It also might mean I'll need to pre-assemble the ingredients for my morning smoothie to allow enough time to jog. The bottom line: building a brain app requires planning ahead, and understanding what resources you'll need to have in place before you can begin your practice regimen. All of this must be factored in and measured against your time constraints.

This pre-planning phase is critical. If you're unable to overcome the inertia in the beginning, your brain will be perfectly willing to let you rollover, push the snooze button, and rely on Future-Self to start tomorrow.

Money/Resources

No matter what field you're hoping to master, securing the right resources is a necessity. Whether it's art supplies, sports equipment, software, musical instruments, books, a coach, or even a chess set, nearly every domain requires some form of additional resources to achieve your goal.

When we met with Chess Grandmaster Susan Polgar at SPICE headquarters, she stressed that one of the exciting things about chess as a pursuit, besides developing the mind, was the low monetary barrier to entry. Chess sets are not expensive, and these days, rather than assembling a massive chess library, aspiring players can view databases with past matches online.

Physical Effort

Fogg points out that the habits requiring significant physical effort can be tougher to achieve. Based on what we know about the brain, its miserly control of glucose, and its tendency toward stasis, this makes perfect sense.

This is where willpower maintenance pays off—never is that trifecta of exercise, nutrition and sleep more important. In Chapter Three, we saw that the Stanford basketball team performed better by lengthening their sleep cycles to combat the drain from extreme physical exertion.

This is also why, when undertaking a physically demanding goal, it's important to understand the exponential and logarithmic aspects of the growth curve and appreciate the power of delayed gratification as you force yourself to expend more energy. Olympic swimmer Michael Phelps, who was notorious for his grueling workouts, capitalized on his rage to master, his coach's creative inventiveness and an astonishing 8000 to 10,000 calories[250] a day to fuel his workout passion in his steady quest for Olympic gold.

Brain Cycles

The amount of mental energy necessary to pursue expertise varies from domain to domain. If not carefully managed, some domains might prove to be overly taxing. Physical and/or mental burnout can lead you to drop a pursuit temporarily or for good.

One of the rock stars of the 17th century's scientific revolution was Sir Isaac Newton. Physicist and mathematician, he is perhaps best remembered for his law of gravitation. As a voracious scholar, his growth mindset caused him to push back against the then geocentric view of the universe and Aristotelian philosophy common at top universities of the time, including his alma mater, Cambridge. Newton was legendary for pouring himself into his work, and juggling a wide variety of academic pursuits at the same time, all the while keeping up a rapid-fire correspondence with some of the greatest thinkers of his era.

Although the basis of his mental stress is not clear, in 1678, he did experience what's commonly described as 'a complete nervous breakdown.' As a result, he withdrew from academia for the next six years, and his correspondence dropped off to a trickle. He eventually returned to his scientific pursuits.

David Bromberg, the multi-instrumentalist who toured extensively and played on many iconic recordings with artists like Bob Dylan, Willie Nelson, and Jerry Garcia, famously burned out and gave up playing publicly for some time, releasing no new music between 1990 and 2007. Again, this is where there is a benefit to investing heavily in willpower maintenance and a clearly articulated plan.

In Chapter Seven we'll explore how meditation can become a cornerstone for keeping your own stress at bay, and willpower at a maximum, as you endeavor toward your long-term goal.

Non-Routine

Relying on habit conserves brain glucose, and of course, initiating new behaviors burns high amounts of it. We can understand why this problem is at the crux of creating a new brain app.

Fogg also says that when you're trying to establish a new habit, you should always be careful to factor this time into your existing routine. Otherwise, when new behaviors intrude on your pre-established schedule, habit acquisition can become difficult.

If you haven't exercised in 15 years and you decide to create a new running habit, it's important to start out with a 'mini step,' such as beginning a walking routine for a couple of weeks before you attempt any jogging. It's also important to figure out how running is going to fit into your day. As discussed, Fogg's research shows that if you gentle your way into a habit, you're far more likely to succeed. The whole idea behind new behavior is setting up the right kind of trigger and then nurturing your motivation and ability with small successfully completed increments.

Going out, buying a pair of running shoes and pushing yourself through a 10k run that afternoon is more than likely to leave you extremely sore, a little defeated and far less devoted to the next 66 days of improvement. Radically attempting to break into a daily routine, even if that routine was largely built on sedentary behavior, carries consequences.

Fogg sums it up this way:

> "Each person has a different simplicity profile. Some people have more time, some people have more money, and some people can invest brain cycles, while others cannot. These factors vary by the individual, but also vary by the context—In studying simplicity, I've found this to be important: Simplicity is a function of a person's scarcest resource, at the moment a behavior is triggered."[252]

In other words, you need to understand the nature of each link in your simplicity profile, and the whole chain is only as strong as its weakest link.

Tiny Habit Method Goes Big

We've built the case for tiny habits and behavioral change on an individual basis, but what happens when a population exponentially multiplies tiny habits? Just how powerful can a habit stack be when you involve an entire dorm floor of college kids?

In the past year alone, Americans used and discarded about 50 billion plastic water bottles. The recycling rate for those same plastic bottles is about 23%. This means that last year, roughly 38 billion water bottles were dumped into landfills or ended up as general litter.[253]

In the face of this growing problem, Elkay Corporation, a U.S company best known for drinking fountains, developed the EZH2O fountain, which doubled as a water bottle filling station. This allows people to reuse their own refillable bottles, thus cutting down on waste.

As a savvy company, Elkay saw an opportunity to make money and be a little greener at the same time. The big question their engineers had was, will people actually change their habits and take advantage of the eco-friendly option? When the engineers finished their filling station design, they decided to add a counter readout on the machine to show users how many plastic bottles they were saving with each fill up.

Once the full expense of research and development are factored in, it's not uncommon for prototype models to scale back from the original concept. "At one point, we almost cut the counter from the specification due to cost," Franco Savoni, VP of Product Marketing and Engineering at Elkay told us. However, the counter stayed in, and here's what happened: in college dormitories across the U.S. where the EZH2O was installed, students got really excited about the bottle counters. They could instantly see a tangible result each time they refilled their own bottle—and they could watch those results stack up.

It took almost no effort, saved them a few bucks a bottle, and provided instant gratification that they were doing something good for society and the planet. Watching the bottle counter turn over each time meant they were, in effect, getting gold stars—small, constant, emotionally satisfying dopamine reinforcers, creating the all important 'tiny wins,' at the heart of Fogg's behavioral model. Fogg makes the point that "emotions create habits."

To up that reward factor, these days students at both University of Michigan and University of Minnesota even organize contests: which dorm floor can save the most bottles and get that counter the highest?

As of this writing, Savoni reports that Elkay has delivered hundreds of thousands of EZH2O units.[254] You can find them in gyms, airports, offices and of course, on college campuses. That's a whole lot of tiny habits built on Fogg's formula of simple trigger, easy ability and small wins to drive motivation.

Fogg says that creating tiny habits is really a design issue. If you're trying to build ecologically friendly machines, or become the next great guitar wizard, or get your travel blog up and running, make sure you've got the basic questions covered. Who, what, when, where, how and why?

In our travel blog example, you would ask yourself specifically who's going to do the writing, what are you going to write about, when and where are you going to write, how are you going to write it, and why does it matter? The more precise you are about your habit intention, the more likely you'll succeed, whether you're writing a travel piece about the Vasa Museum in Stockholm, Sweden, or attempting to make a small dent in cleaning up the planet.

The Role of Emotions

The concept of building a 'mini habit' is not difficult to understand, but there are variables complicating habituation. One is the role of human emotions.

In his book *Self Comes to Mind*, Professor Antonio Damasio describes emotions as complex, largely automatic neural programs of action.[255] He writes that emotions can be triggered by real-time events, events of the past, or related images. They tap into various brain regions, including the areas concerning language, movement, and reasoning. This, in turn, sets off a chain of chemical reactions. So you can understand basic emotions as habit code that has come preprogrammed and pre-installed in your brain at birth.

Certain kinds of emotions tend to activate specific brain regions, producing a kind of lock and key effect. For instance, frightening events unlock the primitive amygdala and trigger additional chemicals associated with fear. When two regions are affected at the same time, it can create a composite or mixed emotion, such as melancholy sweetness or nostalgia.

Feelings are the body's readout of what's happening internally, combined with your moment-by-moment state of mind. As Damasio says:

"Feelings are the consequence of the ultimate emotional process: the composite perception of all that has gone on during emotions—the actions, the ideas, the style with which ideas flow—fast or slow, stuck on an image, or rapidly trading one for another."[256]

During an emotional state, our rapid body readouts allow us to weigh the likelihood of reward and punishment, all in an attempt to predict what might happen next and what we'll do about it. Basic emotions like fear, anger, sadness, and disgust can be understood as a more nuanced approach to the evolutionary choices of fight, flight, or freeze.

Damasio adds that the brain's emotional process follows the same strategy as our body's immune system. When a swarm of outside invaders show up, our white blood cells dispatch an equal number of antibodies. These cells lock onto the surface shapes of the trespassers in an attempt to neutralize them.

Similarly, when you find yourself in an alarming situation, the amygdala dispatches commands to the hypothalamus and the brain stem, increasing your heart rate, blood pressure, respiration pattern, gut contraction, blood vessel contraction, cortisol release, and a metabolic ramp down of digestion, culminating in a contraction of the facial muscles we have learned to interpret as a frightened expression.

In primitive times, depending on the context of the situation, you might freeze in place, where you'd begin to breathe shallowly—important if you're trying to remain motionless in order to elude a predator. On the other hand, you might make a run for it, resulting in an increased heart rate to drive blood into your legs. And your cognitive resources would be redistributed; interest in things like food or sex would temporarily fall by the wayside.

All of this takes a giant toll on your energy reserves. It's costly—especially if it turns out to be a false alarm. Like a dimmer switch, the basic emotions give us graduated options. Instead of entering full on combat mode when an encounter goes poorly, you may choose to simply show your disgust toward that individual, thereby saving precious glucose and decreasing the chance of getting knocked on the head.

Damasio suggests that basic emotions are unlearned, automated, and predictably stable programs resulting from natural selection and genetic predisposition. Although you can choose to 'act' bravely, no amount of stoicism can undo the fear you're 'feeling' on a basic physiological level. This helps explain why people who have performed heroic actions frequently shy away from describing themselves as courageous. Although their outward behavior was brave, they remember feeling profound terror.

It's not a matter of being born with a fearless personality. Firefighters, Navy Seals, and many others undergo intense training to learn how to successfully function through their natural fright and override inborn emotional programming.

Such is the power of intentional habit code. Brain apps allow us to tamp down certain emotions in order to make it successfully through life and death situations as well as piano recitals, company speeches, chess tournaments, and trumpet solos.

Knowing how to override a primitive emotion like fear, a common component of fixed mindset, might have come in pretty handy if you were the CEO's of Blackberry a few years ago.

Motivation

Hope, fear, pleasure, pain, acceptance or rejection—these are key emotions that can come into play with habit change. Because a discussion of emotion can carry so many connotations, B.J. Fogg uses the term "core motivators" to hone in on greater specificity in his behavior model.

When you're attempting some new project like writing a travel blog, you can both be motivated by the hope of thousands of readers, or by the fear that no one will read it. Fogg says you can maintain hope through "success momentum." That is, you can start by working toward—and celebrating—very small wins. This works because the brain has trouble assigning proportionality and will often judge a small success as equally important to a larger one.

In other words, you can knowingly exploit your brain's proportionality bias for your own gain. Remember when you used to get gold stars on your fourth grade spelling quizzes? Taking time to feel good about minor successes still has impact today. Enough small wins can keep you motivated and create a domino effect, where one successful tiny habit can trigger another tiny habit and lead to a powerful and useful habit stack.

Pleasure and pain as emotional readouts are well understood. As we've discussed, pleasurable outcomes and small wins are much more likely to keep us on the habit path. Another way to leverage the emotion of pleasure is to build a reward component into your new habit structure.

Physiologically, rewards help ensure myelination of desired neural connections. Our sensory systems are primed to avoid pain; accidentally touching a red-hot burner on a range top doesn't require a lot of repetition to create a habit of avoidance. The desire to avoid physical pain associated with alcohol and drug abuse is a key element in many addiction recovery plans.

Some habit acquisition is relatively painless. Some is so effortless that we can form habits without even noticing. Still, some involves discomfort—maybe not as severe as burning your fingers, but a noticeable level of mental or physical stress. It's unlikely you have hours and hours of idle time on your hands, which you've spent staring off into middle distance, saving the space for a future habit.

Every new behavior that you seek to create will likely dislodge a pre-existing one. The disruption will not be restricted to your daily schedule. There will also be emotional disruption in which you'll experience new feelings as a result of change. Being aware of, and properly dealing with, both your new feelings and changes in schedule will ensure your habit gets off to a good start.

It's difficult to maintain a steadfast and uncompromising allegiance to an exacting set of behaviors day in and day out. Although we operate and rely on a significant amount of pre-programmed neural code, we are not automatons. As humans, we are infinitely fallible.

Fogg cautions against feeling guilty when you stray from your newly minted habit routine. If you miss doing a tiny habit once in a while, don't beat yourself up over it. If you find yourself feeling guilty and your misses are more than occasional, consider reexamining your habit anchors. He says "if you want to do a habit ten times a day, find an anchor that occurs ten times a day."[257] It's about trial and error, figuring out what anchored behavior you can piggyback on as your trigger.

Fogg says to make sure you take time to celebrate the completion of each tiny habit repetition. It might seem silly at face value, but celebration allows you to leverage pleasure as a way of enhancing habit wiring. Emotional responses heighten and amplify your feelings, which reinforce memory. Fogg's research is clear: the people who celebrate tiny habits are more successful at building them.

Habit Implementation

There are two common ways to think about gearing up for a new habit. One way is to imagine a positive outcome, a fantasy, like standing on the stage, receiving an Oscar for Best Travel Blog Writer as the crowd rises to their feet in standing ovation. The other way is to ponder what about your current situation makes you unhappy and what the solution to your problem might look like.

New York University Professor of Psychology Gabriele Oettingen decided to test these ideas to see which one was more useful. She ran an experiment in which she divided her subjects into three groups and gave them the same problem to solve. The first group was told to start by indulging themselves, fantasizing about having solved the problem. The second group was told to dwell on the negative consequences of not solving the problem.

The third group, however, was told to do both: to envision the satisfaction of solving the problem, and the disappointment at failing to solve the problem.

It turned out that when expectations of success were high, the third group performed best. However, when expectations of success were very low, this same group invested less in planning and backed away from taking responsibility for the outcome.

Oettingen theorized that, as is often the case in business and life, people tend to do a cost-benefit analysis of potential outcomes and gear their actions accordingly. We only invest when we think any idea or a goal has true viability. It appears that we also operate under this same principle when deciding how much energy to devote to building a new brain app.

Motivation is a necessary force, spurring you toward any goal. Countless books and articles tout the benefit of positive thinking as a catalyst, but in Oettingen's book *Rethinking Positive Thinking*, she argues that positive thinking only takes you so far.[258] It's true that fantasizing about achieving a goal can lower blood pressure, compelling us into a state sometimes associated with the desirable effects of meditation. That's because, bottom line, there's a little Walter Mitty in all of us; daydreaming feels good.

Oettingen says, unfortunately, daydreaming can be counterproductive. You feel less urgency to act, because on some level, you've already experienced a fantasized positive outcome. Fantasize too much, and you can find your motivation slipping and reliance on Future-Self growing.

This isn't to say we should eliminate wishful thinking altogether. Oettingen writes that it can be a useful tool, provided you add a few more steps to the habit building process. To increase your chances of success, she's even created a mnemonic device: **WOOP.**

W stands for wishful thinking, which, as we've explained, is a useful starting point. Before you start building a new habit, you want to get excited about what your new habit could mean for you. Fantasizing a desired outcome can be important, but it's only the first step.

O is for outcomes. Oettingen says that, for a well-defined goal, think beyond vague dreams of excellence. The greater the specificity, the easier it will be to develop a concrete plan for wiring a new habit. Dreaming about becoming a long-distance runner is one thing. Aiming to qualify for the Boston Marathon on April 20th 2020—and thus following Hal Higdon's Ultimate Training Guide to hit a time of 3 hours and 40 minutes as a woman age 35-39—is quite another.[259],[260]

The next **O** stands for obstacles. Here is the all-important reality check. By anticipating and examining obstacles, you can get a jump on overcoming issues before they arise.

Oettingen makes it clear that this is not simply about looking at environmental impediments that you might encounter, but more importantly, this is about examining your own *mental* obstacles. These include your heuristics, beliefs and the emotional components that drive your behaviors and mindset.

Chapter Four introduced us to Masaaki Imai and the concept of Kaizen. Central to Kaizen is the idea of solving problems by focusing with laser-like intensity on the root cause. This is precisely what Oettigen is suggesting; move past circumstantial barriers and focus on your own emotional drivers.

Are you acting or not acting out of fear, or shyness or embarrassment, insecurity, or something else, perhaps an overriding fixed mindset, or an outdated heuristic? Oettingen suggests this is the real opportunity to rewire. You need to recognize the root of the particular mental obstacle that's keeping you trapped in your current way of thinking. This requires a careful and thoughtful deep dive into holdover effects of past experiences and understanding how they continue to shape your decisions.

Taken together, the **WOO** of **WOOP** brings us to what Oettingen calls "mental contrasting"— examining the entire scope of the problem, all potential gains and impediments, with a special emphasis on emotional underpinnings and then crafting a *plan* to overcome your obstacle.

This is where the **P** in **WOOP** comes from. Oettingen calls the plan an "implementation intention." A key ingredient to the implementation intention is an if-then statement: 'If I'm afraid to speak in public, then I'll join Toastmasters and learn how to overcome my glossophobia.' The if-then statement reflects both the obstacle and the practical solution.

Oettingen says the real power of Mental Contrasting is examining the mental obstacles that are holding you back and then devising a detailed plan for success.

Pre-rehearsed if-then statements act as defensive plays to combat the urges to procrastinate and delegate to Future-Self. 'If I don't feel like jogging this morning, then I'll crank some hip hop for a few minutes to get myself pumped up before I hit the pavement.' Or: 'If I don't feel like jogging, then I'll start out walking until I build a little momentum and then I'll start my run'.

As you draft your implementation intentions, there are a few things you'll want to remember. Don't link them to times of day—for example, 'If it's 6 a.m., then I'll start jogging'. This has a limiting effect; when the clock ticks to 6:01 am, it's too easy to give yourself permission to lie back and grab a couple more Zs.

If-then statements sync well with BJ Fogg's Tiny Habit Method: 'If I brush my teeth, then I'll floss immediately afterwards'. In addition to keeping you on track when your motivation starts to flag, if-then statements formalize and help lock down exactly what the next action should entail.

It's also possible to piggyback multiple "then" statements on the back of one "if": 'If I enter the weight lifting room in the gym, then I'll do leg lifts, bench presses, and pull-ups'. Or they can be very specific to a practice pattern: 'If I begin doing pull-ups, then I'll take my time and concentrate for five seconds on each contraction to build up my bicep strength'.

Implementation intentions are a great example of seemingly small but potent incremental shifts. They are like micro catalysts to tiny habits. Often, we teeter on the edge of commitment, where the slightest breeze can push us into the ever-waiting arms of Future-Self. Implementation intentions act as a mooring, keeping us secure and steadfast in our habit commitment.

How effective are implementation intentions?

Across 94 studies with over 8,000 subjects in pursuit of a wide variety of habits, implementation intentions outperformed other habit motivation strategies.[261] This might be due to the strong cause and effect linkage implicit in the technique. The 'if' portion of the statement acts as a powerful trigger and the 'then' portion directs a specific action. As we've seen repeatedly, specificity of both trigger and call to action plays an oversized role when you're attempting to implement any goal strategy.

In a 2008 implementation study, researchers looked at 107 table tennis players from German league play: 76 men and 31 women, with an average age of 34.[262] The subjects were split into three groups. The first group served as the control group; they were told to play their matches as they normally would.

The second group was coached to "play each ball with upmost concentration and effort in order to win the match."[263] The third group was instructed to pick out some of the negative thoughts that regularly popped into their heads during match play and shield against these reoccurring mental obstacles by creating appropriate if-then statements.

Group three identified 18 negative thought states that occurred regularly during match play, including feeling anger, exhaustion, and distraction. Subjects in this third group constructed implementation intentions before their next match, like 'if I feel angry, then I will work on relaxing myself', or 'If I feel demotivated, then I'll push myself even harder', or 'If I am playing too cautiously then I'll play with more abandon'. In the next set of matches, group three utilized it's newly created implementation intentions.

After the matches were played, the players' observations reviewed, the coach's critique done, and the scores posted, not surprisingly, group three (the implementation intention group) turned out to have more wins.

The study identified two more bonuses to an implementation intention approach. First, even if you don't have preplanned if-then statements at the ready, they can be formed on the fly once you identify the nature of your mental obstacle. And second, because implementation intentions are so straightforward, it doesn't take a lot of cognitive energy to create them and put them to use.

Implementing if-then statements allows you to plug the dike before emotions can swamp your intention. And it doesn't matter what level of mastery you've achieved in your particular domain; the negative tapes that play in your head turn out to be extremely common.

Even world class scientists, artists, and athletes encounter reoccurring negative monologs. The strategy to offset these negative feedback loops doesn't have to be sophisticated. The key is to recognize negative mental feedback and counter with the proper if-then statement before your analytical System 2 begins its all too common process of second-guessing your behavior.

Habit Tracking

Consistency and simplicity are two keys to keep in mind when attempting to maintain and track new habits. For consistency, we'll turn to an unlikely source: a piece of lore surrounding comedian Jerry Seinfeld.

Love it or hate it, Seinfeld's sitcom "about nothing" reportedly managed to earn him a cool $267 million in 1998 alone, according to writer James Clear,[264] turning him into a comedic icon.

For instance, perhaps you've heard of the Seinfeld Productivity Program?

Clear quotes software developer and aspiring comedian Brad Isaac about a chance encounter in a comedy club, wherein Isaac had a chance to ask advice from the comedy legend and future animated bee himself:

"He said the way to be a better comic was to create better jokes and the way to create better jokes was to write every day.

"He told me to get a big wall calendar that has a whole year on one page and hang it on a prominent wall. The next step was to get a big red magic marker. He said for each day that I do my task of writing; I get to put a big red X over that day.

"'After a few days you'll have a chain. Just keep at it and the chain will grow longer every day. You'll like seeing that chain, especially when you get a few weeks under your belt. Your only job is to not break the chain.'"

It's an appealing story. There is just one problem. It never happened.

As James Clear himself acknowledges, Seinfeld denied the incident in a 2014 Reddit thread. "This is hilarious to me," Seinfeld wrote, "that somehow I am getting credit for making an X on a calendar with the Seinfeld productivity program. It's the dumbest non-idea that was not mine, but somehow I'm getting credit for it."[265]

In fairness to those still telling the tale, it's not the first time a well-known idea has been misattributed.

Besides, neither detail necessarily means that the method itself is without merit. It is true that persistence is a major key to success and it is important to commit to a regular deliberate practice schedule, especially in the early incubation period of a brain app.

This brings us to the second idea: simplicity. The simpler the plan for staying the course, the less brain glucose is expended. As we've seen in previous chapters, the brain tends to be downright miserly with its fuel. This is where the principle of Occam's Razor comes in.

Occam's Razor is a philosophical rule of thumb. Sir William Hamilton coined the term in the 19th century, cashing in on the celebrity status of 14th century scholar, logician, theologian and friar William Occam.[266] The friar was long dead, and thus not in a position to explain that he didn't have much to do with his supposed Razor. Occam's Razor suggests whenever you're constructing a mathematical solution, you should reduce your chance of error by eliminating any redundancies. Crudely put, the simplest solution is the best.

The idea of a law of economy or parsimony is not new. "We consider it a good principle to explain the phenomena by the simplest hypothesis possible," said Egyptian astronomer Ptolemy around 1800 years ago.[267] This concept is present in the works of Aristotle and Thomas Aquinas, as well as many other legends of philosophy.

In the end, it shouldn't really matter where an idea or strategy comes from as long as the content and intention is fair and sound. If your goal is to create quality brain apps, you might consider the amalgamation of two ideas, consistency and simplicity— we've taken the liberty of calling "Seinfeld's Razor," as a reminder to maintain and measure your practice regimen. We've already alluded to this earlier.

Try putting an X on the calendar for every day you manage to hang in there on your new habit goal. Remember, a little positive affirmation will help reinforce the links of your habit chain.

Depending on the brain app you're trying to build, the timing will vary, but on average, 66 days of X's will be a substantial down payment on a new piece of intentional programming that can move you one step closer to your long-term goal.

Chapter Key Points: Hacking Your Habit App

- Almost half of our everyday activity is driven by pre-programmed neural wiring commonly know as habit.

- Some of these brain apps came wired at birth and some are wired unintentionally through repetition.

- Heuristics are a mental shortcut, a framework for making decisions. If you want to understand yourself better, identify and examine your heuristics.

- BJ Fogg's Tiny Habit Method involves taking the time to identify the right behavioral trigger, and piggybacking on an existing habit to create a powerful habit stack.

- Keep in mind that momentum is built out of tiny wins.

- Consider your simplicity profile, understanding your strengths and weaknesses with regards to time, money/resources, physical effort, brain cycles, and the effects of non-routine.

- When you're stymied by a lack of progress, remember psychologist Gabriele Oettingen's mental contrasting model. Use her system of **WOOP**—**W**ish, **O**utcome, **O**bstacle and **P**lan—to examine your current mindset.

- To prepare yourself against Future-Self and the dreaded Extinction Burst, take some time before you embark on a challenge to devise some if-then statements.

Chapter 6
Creativity App

We now know the essentials of what it takes to build a brain app. To take you to the next level, you'll need to harness your creativity. You might think of creativity as the realm of artists and musicians, but the truth is, mastering nearly any skill requires you to think independently and find novel solutions to problems as they arise in pursuit of your long-term goal.

Not only does creative problem solving allow you to avoid *stagnation swamp*, but it's fundamental to boosting application strength, which compounds your chances of success.

So just how does creativity work? Only recently has neuroscience begun to answer that question. I say "begun to answer" because understanding creativity is still very much a moving target. Let's start with what we do know, by unpacking some common misconceptions.

Left Brain/Right Brain

By now, you're familiar with the idea that there are two hemispheres in your brain connected by a corpus callosum. Undoubtedly, you've also heard that the left hemisphere is responsible for analytical thinking, while your right hemisphere is the home of creativity.

It's probably less likely you've heard of Michael Gazzaniga. However, he is one of the founding fathers of cognitive neuroscience. Early in his career, he teamed up with famous brain pioneer Roger Sperry. Sperry shared the 1981 Nobel Prize, with David Hunter Hubel and Torsten Nils Wiesel, for their work in split-brain research.[268]

Gazzaniga's own split-brain work involved patients in treatment for severe epilepsy who underwent operations severing the bridge between their brain hemispheres.[269] Cutting this corpus callosum was an attempt to isolate and contain the epilepsy so it couldn't spread to both sides of the brain. His studies led to the discovery of what's called the functional lateralization of the brain: how the two hemispheres can operate independently and still communicate with each other.

In the mid-sixties, the media seized on these findings and ran with them, morphing the data into a still-familiar narrative of the artistic, creative right brain and the methodical, analytical left brain. But although you can find some degree of specialty, there's also a tremendous amount of overlap and redundancy between the hemispheres. The right brain versus left-brain model is at best, a tremendous oversimplification and at worse, just plain wrong.

In the days since Gazzaniga's groundbreaking lateralization research,[270] numerous fMRI studies have searched for the locus of creativity. However, creative thought requires the interplay of so many different systems that trying to find where creativity happens in the brain is a little like visiting an automobile factory in the hopes of pinpointing the exact spot where cars are created. Should we look to the assembly line, or paint line, parts department, engineering department, or front office? The truth is, it's all of those places working in conjunction with each other.

It takes multiple neural networks to develop an original thought. Unfortunately for our purposes, this means there isn't a single roadmap to creativity, just as there is no single way to get to Chicago. Instead, there are many transportation modes and numerous routes.

Another complication in any study of creativity: by its very nature, it can be difficult to measure. How do you take something so subjective and convert it into clean, numerical data? Various people have tried, with decidedly mixed results.

Creativity is, at least for now, a partially hidden enterprise because parts of our brain are not yet accessible to scientific scrutiny. We have made gains, but the ability to replicate creativity remains elusive.

This is why it's impossible to excel at trumpet exactly like Dave Guy or at chess exactly like Susan Polgar. Doing so would require you to have lived their identical lives, down to the nano-second, with every bit of their realized experience and genetic coding unchanged. By that, I mean you would actually have to be them; even cloning wouldn't work because of a little variable in human brains called transposons.

Transposons

A transposon is a fragment of DNA that inserts itself into another cell. Research suggests that about half our DNA sequence is made up of these fragments, these interlopers.

In the cells of, say, your lungs, heart, or kidneys, transposons have no real effect. They don't behave like viruses, which sneak into cells and multiply like crazy. They're more like very mellow hitchhikers; once they've found their way in, they're usually content to fall asleep and enjoy the ride.

The exception is the brain. Once transposons get inside neurons, they can alter the very nature of the cell. It's like a troupe of improv actors that show up unexpectedly at your birthday party; suddenly, you're at a very different party. Transposons can influence a neuron's firing sequence, turn it off or on, or even reconfigure the operating code of the whole chemical-electrical switch.

This means they can change the entire identity and purpose of a neuron. And like any good improv actor, they can shift into a variety of roles and characters.

Kelly Clancy reports that it's a Darwinian parable playing out on a cellular level,[271] producing "a kind of evolution in miniature."

The result? Even among twins, no two brains are exactly alike. Identical twins begin with identical DNA, but the arrival of those improvising transposons makes neural activity wholly unique. And since transposons aren't passed down, your brain is truly a once-in-a-lifetime show.

If all this sounds a little, well, scary (rogue DNA wreaks havoc on unsuspecting brain!), think of it this way: transposons, in essence, help to make each of us the completely unique creative individuals we are, from Dave Guy to Susan Polgar to you.

IQ and Creativity

For a long time, conventional wisdom held that if you wanted to compare one person's mental powers to another's, you didn't need to look any further than their respective IQ scores. So, were the Fords, Edisons and Jobs at a creative advantage because of superior IQ? What is the relationship between creativity and IQ? The modern day Intelligence Quotient test, more commonly known as the IQ test, analyzes your applied knowledge of math, as well as verbal and spatial recognition skills. At the end of the test, your final score is reflected by a number between 1 and 200, with 100 being average.

We can trace the American fixation on IQ back to the beginning of our involvement in World War One. The U.S. War Department was searching for ways to identify who would be best suited for which jobs, from foot soldiers to officers. For help, the military turned to psychologists like Lewis Terman of Stanford University.

Terman had tweaked an intelligence test devised by the famed French psychologist Alfred Binet to create a new version, called the Stanford-Binet Intelligence Scales.[272] Binet initially promoted his model as a tool for classifying developmentally disabled children, but the U.S. military was so impressed with Terman's version, they hired him and six others to create the "Army Alpha", an assessment test which was administered to 1.7 million GIs.[273]

At the time, there was no other widely circulated intelligence test to use as a benchmark, so it's hard to measure the test's net effect. However, the allies went on to win the war, and after the war, Terman went on to screen children for signs of "genius level" IQ.

Several years later, Terman used these screening results to kick off a study aimed at understanding the wide-ranging effects of "genius." (Eventually, the study would abandon the emotionally charged—and difficult to quantify—label "genius" in favor of "gifted.") He began in 1921 at Stanford. Terman looked at 1,500 boys and girls, attempting to track factors like developmental progress, their interests when playing, their medical condition, how much they read, and how many books were available to them at home.[274] Then he continued to periodically check in with those same subjects throughout their lives.

An early example of a longitudinal study, it's also the longest running of its kind and still continuing today, to be concluded at the death of its final subject.[275]

This work eventually begat Terman's multivolume *Genetic Studies of Genius*,[276] considered a seminal document in American psychology. (That's not to say that Terman's scholarship all holds up by today's standards. In testing across cultural and racial groups, he jumped to a number of sloppy, prejudiced conclusions.)

However, he did debunk then-common misconceptions about high-IQ children: his research did not show them to be physically frail or socially maladjusted. In an era where parents often held their children back a grade to prevent their child from being the youngest in their class, Terman found that being the youngest in a class was, in fact, a predictor of a high IQ.

For our purposes, Terman's most interesting result concerns creativity.

To the extent it could be measured, Terman found that the 1,500 high IQ study subjects did make an above-average number of societal contributions in creative fields. This, on the face of it, would suggest that a high IQ delivers a key creative boost. However, in a separate study, sociologist Pitirim Sorokin showed that a random group of children from equivalent socio-economic backgrounds would do just as well.[277]

In other words, Sorokin's work suggested that a high IQ score was not itself a predictor of creative success; rather, children growing up with above-average access to certain resources—books, piano lessons, etc., and the undivided attention of a parent or parents who didn't need to work multiple jobs—had advantages over the population at large.

Terman's study arguably had other shortcomings. Luis Alvarez, a San Francisco native who later became a Nobel laureate in physics,[278] did not meet Terman's definition of genius as a child and wasn't included in the study. William Shockley, who was a 12-year-old Palo Alto resident in 1922, also failed to make the 'genius' cut, even though he would go on to share a Nobel Prize in physics with John Bardeen and Walter Houser for the invention of the transistor.[279]

Interestingly, Shockley credited his team's success, not with their collective IQs, but with a trial-and-error approach he described as "creative-failure methodology."[280] It was more a matter of perseverance than pixie dust. "A basic truth that the history of the creation of the transistor reveals," he wrote, "is that the foundations of transistor electronics were created by making errors and following hunches that failed to give what was expected."[281]

Due to his role in commercializing the transistor, Shockley is now considered to be one of the founding fathers of Silicon Valley.[282] (Another founding father, coincidently, was Lewis Terman's own son, Fredrick.[283])

Of the children in Terman's study who did exhibit 'genius' level IQs, roughly a third of both the men and women did not graduate from college. Faced with the data, even Terman had to admit: "We have seen that intellect and achievement are far from perfectly correlated."[284] This sentiment is echoed by many other studies; abnormally high IQ is no guarantee of academic success or high creative output.[285]

It's been suggested that Terman's research supports what's called the Threshold Theory, which states that an IQ of 120 is enough to achieve "creative genius".[286] Anything above that point doesn't seem to make much of a difference. Since the average IQ is measured at 100, this would suggest that creativity is well within the reach of a large number of people, given hard work, focus, and a certain dose of luck.

But in 1983, the story got more complicated. Howard Gardner published his groundbreaking book *Frame of Mind*[287] and turned the notion of IQ upside down.

Gardner argued that the standard IQ test doesn't give us nearly the whole picture of intelligence. How can one number possibly give us any useful information about something as complex as a human mind? He suggested we were missing a whole boatload of emotionally driven indicators, which had a profound effect on our intellectual process. Collectively, these additional factors were first called Emotional Intelligence by researchers Peter Salovey and John D. Mayer.[288] The idea became more widely known with Daniel Goleman's international bestseller *Emotional Intelligence: Why it can matter more than IQ*.[289]

Gardner says that we navigate the world and weigh our decisions using all manner of domain knowledge that exists outside the confines of test parameters, along with emotional drivers, including mind frame, self-awareness, relationship management and self management.

So if standard IQ tests don't tell the whole story, and high IQ isn't a solid predictor for creativity, what about tests that are specifically designed to measure creativity?

Divergent Thinking

As we've noted, scientists trying to get to the core of creativity encounter a very basic problem at the outset. Unlike, say, size, or time keeping, creativity is extremely difficult to measure and test. It's even difficult to define.

Frequently, people searching for a creativity metric focus on what's called "divergent thinking". This is the ability to come up with a large number of solutions to a given problem. It's the 'no wrong answers' school of brainstorming. Divergent thinking is all about casting the widest possible net, and then gauging success from overall net size.

There are arguably some benefits to this approach. Since divergent thinking concerns itself with the sheer amount of ideas generated, measuring it is as simple as counting.

What do these experiments look like? Imagine that you're handed a paper cup and asked to think of as many uses for that cup as you can. A person with a knack for divergent thinking would be off and running:

a drinking vessel, a flycatcher, a drain stop, a place to store crayons, a hat, and so on. One of the standard divergent testing questions is, "How many uses can you devise for a brick?"[290]

This approach lets scientists easily assign scores to large groups of test subjects, generating huge amounts of easy-to-interpret data. Assuming, of course, that divergent thinking is a useful lens for examining creativity in the first place. If you've ever walked out of a 'no wrong answers' brainstorming session feeling unsatisfied, you may already grasp the controversy at play here.

Some scientists dismiss the relevance of divergent thinking, arguing that, at the very least, it's not a useful way to assess a person's creativity. Sure, it's easy to compare one person's score to another, but it's difficult to prove that high scorers in the test environment are more creative in real-life situations.

For one thing, divergent thinking tests don't seem to have any correlation with a person's future creativity. Some high scorers might just be really good at playing the divergent game, in the same way that you might be a regular J.P. Morgan when it comes to Monopoly, but that alone doesn't qualify you to run a real-life corporation.

There isn't much evidence that finding many uses for a brick on any given day translates into any creative advantage later in life. And outside of a testing facility, most of the time, solutions only count as solutions if they're actually useful. In other words, a paper cup would make a terrible hat.

In addition, most people would agree that when we judge a person's creative output, quality trumps quantity. Originality or novelty is considered an essential part of the mix. Judging a person's creativity only by their number of ideas is like saying that romance novel writer Nora Roberts, who has published a massive amount of books—over 200—is a more creative writer than legendary poet and author Maya Angelou.

Furthermore, creativity seems to be domain specific with a direct correlation to domain knowledge and not a generalized trait that applies across the board, as divergent thinking tests would suggest.[291]

To add another wrinkle, University of Iowa neuroscientist Nancy C. Andreasen suggests that the human race might owe far more of its creative achievements to convergent thinking, the direct opposite of the divergent approach.

Convergent thinking doesn't concern itself with finding a lot of answers, but with winnowing down to the single best solution. "A process," she notes in an article in *The Atlantic*, "that led to Newton's recognition of the physical formulae underlying gravity, and Einstein's recognition that $E=mc^2$."[292] But nobody is clamoring to test for convergent thinking. It's tough to know just how to tally it.

It appears that creativity isn't lightning bolts hurled down by Zeus. And it doesn't seem to flow from brick ideation or from the netherworlds of our individual DNA as represented by super high IQ.

Top Down Process

What does it look like when your brain grapples with a creative problem?

Let's turn back to psychologist Daniel Kahneman. Kahneman says problem solving engages System 2 as part of a top-down analytical process, which involves your prefrontal cortex, the home of executive control. System 2 comes online when you deliberately focus on a problem, applying conscious thought to it. Your prefrontal cortex begins to run through a checklist of possibilities, searching out appropriate matching patterns or solutions. And where does your brain look for those answers? In the library of past experiences we call memory.

With the sum of your accumulated knowledge as a guide, the top-down creative mode allows you to form a specific hypothesis and test it. The scientific method is an example of this approach.

Neuroscientist Arne Dietrich says the beauty of this kind of empirical process allowed the NASA engineers to bring the crippled Apollo 13 spacecraft's astronauts home safely, allowed Bach to compose hundreds of cantatas, and Thomas Edison to methodically test all of the many possibilities that led to the incandescent bulb.[293]

If it all works out, you find an appropriate solution. Sometimes, as with the case of Edison and the incandescent light bulb, the process can drag on for an exhausting 14 months of experimentation.[294]

Scaffolding

Although some problems can be difficult to solve, the brain's ability to project into the future and run multiple simulations means we can fast forward possible outcomes and avoid potential dead ends. We can imagine an eventual outcome or long-term goal and work toward it.

The best-selling novelist and Academy Award-winning screenwriter John Irving says in the introduction of his book, *A Prayer for Owen Meany*, that his habit is to write the last sentence of a new novel first.[295] Some neuroscientists call this *scaffolding:* a problem-solving strategy in which you identify the end point and then fill in the missing steps, making sure the steps sync with the final goal.[296]

This can be a real advantage. A known or hypothesized conclusion makes it easier to connect starting point A to endpoint B. It also allows for task sets, which are the pre-identified parameters for what will and won't be included in the search. Again, this has the effect of narrowing the search and creating more focus on the necessary steps for achieving a logical end goal.

The downside with this kind of creative processing is that narrowing your search parameters ahead of time can shut out some unusual, surprising solutions.

Take, for example, clostridium difficile, the gut disease commonly known as C. diff. This contagion can cause maladies ranging from unpleasant digestive problems and diarrhea to life-threatening symptoms like severe inflammation of the colon. Worse, in recent years, particularly in hospitals, it's become a growing problem, both in frequency and severity.[297]

A successful treatment for eliminating C. diff is to transplant a fecal sample from a healthy individual into the infected patient's gut. That's right, I said "fecal sample". The antibodies from the transplanted fecal matter settle into the patient's gut biome, multiply, and then overpower and wipe out the C. diff bacterium.

One of the problems with the novelty of fecal transplant is the idea of ingesting someone else's feces, no matter the delivery method. It's probably not the first treatment that comes to mind when you're trying to combat a gut disease. Employing a typical top-down process to combat C. diff would likely eliminate this approach from consideration. This is where analytical, creative problem solving can, unfortunately, blind you to counterintuitive solutions.

Luckily, your brain has a backup plan.

Bottom Up Process

This brings us to the second kind of problem solving, the one associated with your unconscious thinking process: epiphanies. It's what many people consider to be the gold standard for true creativity.

The process starts when your more rational, top-down, conscious thinking System 2 struggles with a question it can't seem to solve. With possibilities exhausted, you start to lose focus. Your attention shifts. Your executive control system bows out, handing over the problem to an unconscious neural network, which turns to memory and the associative matching task. It's like an architect who, out of ideas, kicks a problem down to the site foreman, saying, 'See what your guys can do with this design issue.'

However, with your unconscious brain region heading up the search, there is no longer any direction or intervention from the executive control center of your prefrontal cortex. Task sets loosen. Associations connect less powerfully and more randomly. This might sound like it would hamper the whole search enterprise, but without the scrutiny and self-criticism of top-down control, the bottom-up process is allowed to operate unimpeded. Wilder, more unexpected answers begin to surface.[298]

In these kinds of situations, random associations are not only free to form but allowed to take root, and it only takes a few to create a new ad slogan, a novel solution to C. diff, or a new concept in architectural design.

It's important to keep in mind that epiphanies are not some bold strike of thought lightning. They represent a whole lot of toil taking place in the brain's basement, by a host of unrecognized worker neurons who struggle away at discarded problems for minutes, hours, days or even weeks at a time without the analytic tools of executive control. Sometimes, even your unconscious can't solve the problem, but sometimes, we are tickled by the answer.

While your prefrontal cortex is on hiatus, should you chance to encounter some new stimulus—like a tub of water you displace by plopping into it—you, like ancient Greek mathematician Archimedes,[299] might spot new associations bubbling up unexpectedly.

I'm talking about that "Eureka!" moment when associative circuits eventually connect, and the hard work that's been going on in the basement of long-term memory is spotlighted in your working memory. That newly-minted idea, having made it to consciousness, can now bask in the glow of the light of day. You're struck with what feels like a momentary insight. If you're Archimedes, you discover a new principle, which turns out to be a central tenant of physics about how fluid mechanics operate.

Today, using fMRI technology, neuroscientists can watch the 'Aha' revelation unfold on a cellular level. Neurons begin to cluster and activity speeds up, eventually giving way to a burst of energy not unlike a mini fireworks show. All this can be witnessed by the fMRI technician about eight seconds before the subject is aware of their impending moment of truth.[300]

So what are insights?

Insights are not merely the rediscovery of misplaced data, like suddenly remembering where your car keys are. They are combinations or reinterpretations of information in the basement of your associative memory at work, creating something entirely different or new. They are the embodiment of what it means to 'think outside the box.'

Take this classic riddle:

> A father and his son are in a car accident. The father dies at the scene and the son is rushed to the hospital. At the hospital, the surgeon takes one look at the boy and says, "I can't operate on this child, he's my son."[301]

How can this be?

This 1970's era brainteaser played on the fact that some readers will automatically assume the surgeon is male. (These days, with an increased awareness of adoption and blended families, there are arguably several solutions to this riddle. However, the original intended answer is that the surgeon is the boy's mother.)

Suppose you are one of those people who is unable to find an explanation right away. It's true your prefrontal cortex might be stumped. Unbeknownst to you, even though some lingering gender bias has brought your prefrontal cortex to an impasse, unconscious neural circuits are triggered as a back-up plan to find the answer. This allows your conscious mind to head off in some new direction.

In that case, all this association business continues below your awareness. When the solution finally floats, fully formed, into your working memory, it feels as if it came out of nowhere, an epiphany. (Epiphany comes from the Greek *epiphainein,* which means to 'reveal.')[302]

Only the conscious brain region has language. This is probably a good thing, because if your unconscious brain region could talk, it might very well demand a thank you for all of its hard work, or at the very least an "I told you so.

Mind Wandering

The ability of reflexive System 1 to perform a mundane task while your mind essentially goes into improvisational mode has a name: mind wandering. What we call mind wandering is actually your freewheeling unconscious engine hard at work, checking out random associations for a winner.

Most people think they mind wander about 10% of the time. Researchers at UC Santa Barbara put that figure closer to 30%. When engaged in well-rehearsed tasks, like driving a car on a wide-open highway, it's estimated that mind wandering can be as high as 70%.[303]

Behavioral psychologist Susan Weinschenk[304] makes the important distinction between mind wandering and daydreaming. According to Weinschenk, daydreaming involves an aspect of fantasy, like imagining you've been asked to headline in the next Star Wars movie, or that you've just won the lottery.

Mind wandering occurs when your brain is engaged in a habituated activity, like driving, and at the same time you're wrestling with some other problem. Mind wandering, according to researchers at UC Santa Barbara, is tied to the associative connections of creativity. Weinschenk notes that the ability to perform a rote task while mind wandering and, more specifically, to switch on this mental meandering at will is "the hallmark of the most creative people."[305]

Thought Experiments

There are numerous stories of great thinkers engaging in some very productive mind wandering forays.

When Einstein was in his early teens, he received a gift from a family friend: a series of illustrated science books with the catchy title of *Naturwissenschaftliche Volksbucher* ("People's Books on Natural Science"), by Aaron Bernstein.[306] In *Einstein: His Life and Universe*, Walter Isaacson quotes Einstein as having later described it as "a work which I read with breathless attention."[307]

In the first volume of Bernstein's popular science series, he asked the reader to imagine a bullet shot through the window of a fast-moving train. Bernstein postulated that anyone examining the bullet's exit on the opposite side of the train would conclude the bullet must have been shot at an angle.

Bernstein's point was that, because the earth is hurtling through space, light would exhibit the same refracting properties going through a telescope lens as the bullet passing through the train windows. And that this outcome would always be the same regardless how fast the source of the light was traveling.

Bernstein wrote, "Since each kind of light proves to be exactly of the same speed, the law of the speed of light can well be called the most general of all nature's laws." (I think we can all agree Einstein went on to do a pretty good job chasing down this idea.)

In a later volume, Bernstein had his readers imagine the effects of traveling through space as a passenger on a wave of light. At sixteen, the young science nerd Einstein was fascinated by these creative challenges. In retrospect, we can see the seeds of Einstein's famous 'thought experiments,' where he meditated on complicated physics problems through striking visualization.

This was partly out of necessity, given the limitations of turn-of-the century technologies. It was difficult to conduct literal experiments in the burgeoning field of physics prior to nuclear accelerators. Today, manipulating a solar system under lab conditions still remains a tall order for even the brightest and most determined physicist.

Taking advantage of the basal ganglia's ability to run simulations, and the brain's inherent visual strengths (Thirty percent of brain activity appears to be devoted to decoding images[308]), Einstein eventually found himself applying Bernstein's approach to some of the toughest physics problems of his day.

In essence, this allowed Einstein to watch a movie version of the physics problem as it played out in his head. And the beauty of this technique was that he had the power to edit, readjust, and rerun the footage over and over as he sought to uncover the underlying principles of space and time. Before the days of particle accelerators and NASA telescopes, Einstein was already making use of the best tools nature had to offer.

There was, however, one small problem: his thought experiments might go off without a hitch inside his own mind, but he still had to demonstrate the results to the greater scientific community.

To prove one of the tenets of his Theory of General Relativity—that light bends when it passes near a very heavy body—Einstein eschewed the lab for a more creative solution: a total solar eclipse. When the moon passes in front of the sun, the moon shields some of the sun's intensity, allowing us to observe distant stars, and to measure their light for refraction. Luckily for Einstein, in May of 1919, English astronomer Arthur Eddington agreed to travel to the Island of Principe off the west coast of Africa to take telescopic photographs of stars during an upcoming eclipse.[309]

From Einstein's point of view, Eddington's task couldn't have been more simple: to demonstrate that starlight would be bent by the warped space around the sun's mass. In reality, given the weather conditions, Eddington risked his life and was barely able to snag the needed photographic evidence. But Eddington got the shots, and his subsequent calculations proved Einstein's theory and helped turn him into the scientific rock star he still is today, nearly 100 years later.

Mind wandering, or what Einstein self-described above as 'combinatory play,'[310] is essentially putting your mind in a relaxed mode where executive control and working memory restraints are loosened, but not completely undone. This allows you to access somewhat more random connections while still operating within reason. In this regard, mind wandering can be thought of as combinational, a hybrid of both top down and bottom up processing, both part of analytical System 2.[311]

Whether you're working top down or bottom up, you're harnessing your solution mechanism. Creative problem solving, even for an Einstein, is not some magical technique wholly different from the way the human brain solves any kind of problem.

There is no secret sauce involved in creative solutions. If you want to improve the number of creative solutions you produce, you must invest time in the process. You'll need to build a brain app to practice, habituate and improve problem solving. And all of this will mean ratcheting up your associative domain knowledge.

Domain Knowledge

What is the most practical method for generating fresh new ideas? Despite our slowly growing understanding of the neural processes governing creativity, that question haunts every scientist, writer, artist, filmmaker, athlete and inventor, or anyone serious about upping their creative chops.

Much has been written about the connection between creativity and what's called the default network, the mental mechanisms associated with mind wandering and daydreaming we introduced in Chapter Two. This might be because there is something inherently alluring about the idea of epiphany, that bolt-out-of-the-blue inspiration seen as the mind fruit of spontaneous genius.

What doesn't get as much airtime is the role of domain knowledge in creativity—that is, the breadth and depth of your familiarity with a given field, including the contents of your associative memory and the sum total of your practice regimen.

One of the best-known epiphany stories occurred around 1666 when Sir Isaac Newton was allegedly conked on the head by a falling apple, thus triggering a revelation on the nature of gravity. Here's some slightly lesser-known context: by 1666, Newton was a master of Euclidean geometry, algebra, and Cartesian coordinates. He'd already invented calculus, which he needed so he could measure planetary orbits.[312]

"In other words," writes Nancy C. Andreasen in 'Secrets of the Creative Brain,' "Newton's formulation of the concept of gravity took more than 20 years and included multiple components: preparation, incubation, inspiration..."[313]

Newton might not have been the first person to be beaned on the skull by an apple. But it's possible he was the first person whose entire career had prepared him to fully grasp the principles behind his 'apple moment.'

Similarly, Einstein's famous thought experiments were more than idle fancies; they were grounded in his expertise in physics. There's a reason Darwin's breakthroughs came in biology and not, say, dance. It's the same reason rock icon Prince was able to bring us his mega hit "Purple Rain" in 1984 at the age of 26 when he mixed pop, rock, and gospel. Prince, the son of a musician, had been playing piano and composing since he was seven years old.[314]

There's no way around it: your best bet for creative achievement is doing your homework. Having an enormous library of information and expertise helps build out the novel connections between ideas that we call creativity. It doesn't guarantee that great works will necessarily follow, but for those of us who enjoy the experience of trying to bring something new into the world, it appears that stocking and constantly restocking your brain's memory library is a prerequisite.

So was Steve Jobs really a creative genius? The visionary techno-wizard of Palo Alto whose own 'apple moment', came with a wave of his hand, summoning from thin air the Mac, iPhone, iPod, and iPad? Or, instead, was he an extraordinary designer, and marketer in a field he was passionate about, who surrounded himself with like-minded experts, and lived out Daniel Kahneman's formula for success (hard work + luck)?

This rewrite on the classic headline, 'Genius has epiphany,' may not sell as many papers or computers. A bolt from the blue is much catchier than a bolt from decades of careful groundwork. But in the end, a lack of pixie dust doesn't diminish what Newton, or any other domain master, has brought us.

Copy, Transform and Combine

In his four-part documentary series "Everything is a Remix," Kirby Ferguson has something to say about the nature of creativity and it doesn't involve pixie dust. "Creativity isn't magic," says Ferguson. "It happens by applying ordinary tools of thought to existing materials."[315]

His central premise is that ideas don't originate in a vacuum. Instead, most creators, from George Lucas to DJ Danger Mouse, borrow and adapt their ideas from other sources.[316]

Artists, inventors, and serious thinkers immerse themselves in their chosen subject, Ferguson explains. Often, the early learning phase involves outright copying others—an effective way of practicing and developing one's abilities. Ferguson reminds us that Richard Pryor's earliest standup was an obvious riff on Bill Cosby's work.[317] Similarly, Bob Dylan began his career as something akin to a Woody Guthrie impersonator. Once would-be creators have assembled a deep domain knowledge and refined their skills, they can start to make their work their own.

Ferguson sums up the process with a simple formula: copy, transform, and combine. Creative people pick up on, or copy, aspects of other material that interest them. Next, the creator elaborates on this idea, tinkering with it and altering it until it is often distinctly different from the original. This element is then combined with other copied and transformed ideas, until a whole work is created. If the tinkering is extensive enough, and the recombinations novel enough, the end result can serve to inspire the next wave of creators.

Ferguson calls these inspired, adapted works "remixes", a word he borrows from hip-hop culture. Hip-hop, which has an established tradition of recombining and reworking samples of older songs, has understood "copy, transform, combine" from the very beginning. Sugarhill Gang's landmark "Rapper's Delight" borrows its bass riff wholesale from Chic's "Good Times," a riff which has since been sampled dozens of times, by everyone from Grandmaster Flash to Will Smith to Daft Punk.[318]

In fact, Ferguson makes the case that most entertainment is a kind of legal remix. Genres like horror, action, or romantic comedy can only exist because writers have borrowed from each other to create a commonly understood palette of ingredients, from jump scares to car chases to meet-cutes. This isn't even getting into the number of films each year that are direct remakes, sequels, or adaptations of pre-existing stories. (We're looking at you, superhero movies).

Ferguson vividly demonstrates how even game-changers like filmmaker George Lucas weren't working from scratch. The iconic movie *Star Wars* seemed to invent a new genre out of whole cloth, but, in fact, much of its power comes from the bafflingly wide assortment of older ideas Lucas threw together, from old Flash Gordon flicks to Westerns to Joseph Campbell's philosophies of myths and heroes. In fact, the original trilogy includes many shot-for-shot copies of scenes from other films, including Akira Kurosawa's samurai movies. (Luke's white tunic in the first *Star Wars* is familiar to any young martial arts student, and his light saber is just a sword plus a laser.)[319]

In 2016, Leonardo Dicaprio won an Oscar for Best Actor in Alejandro González Iñárritu's *The Revenant*. Much like Star Wars, the film includes some striking visual parallels to another director's work, namely Soviet filmmaker Andrei Tarkovsky. Iñárritu openly admits to Tarkovsky's influence, among others.[320]

Triumph of the Remix

The key to novel recombination is making surprising new connections across a wide variety of disciplines. Which brings us to Tony-winning composer, actor, rapper, and MacArthur "Genius" winner Lin-Manuel Miranda.

While reading Robert Chernow's biography of Alexander Hamilton, Miranda noticed striking parallels between the life of America's first treasury secretary and the classic hip-hop narrative. The result was *Hamilton*, a hip-hop biographical musical narrated by Aaron Burr, Hamilton's political rival and eventual killer. It's a fresh take on the monumental brilliance and equally monumental egos of our founding fathers, capturing the 18th century revolutionary spirit for new generations of Americans. It is also a stunning example of a Kirby Ferguson-style remix.

The beauty of *Hamilton* is not just the novelty of Miranda's recombination—on the face of it, a rap show about a long-dead treasury secretary doesn't seem like a guaranteed success. But Miranda's extensive domain knowledge of both hip-hop and musical theater, combined with years of historical research and careful tinkering, allow him to translate the story of America's early days into a modern, effective format.

For instance, he captures the machismo and illicit danger of eighteenth century dueling with "The Ten Duel Commandments," an obvious homage to "Ten Crack Commandments" from rapper Biggie Smalls. But even beyond Chernow and Biggie, Miranda cites an almost George Lucas-like array of inspirations, including musicals like *Evita* and *Les Misérables*, the immigrant experience of Miranda's own father (who, like Hamilton, was born in the Caribbean and arrived in America as an ambitious, driven teenager), the free-wheeling comedy podcast *My Brother My Brother and Me*, Aaron Sorkin's *The West Wing*, and the insightful criticism of Miranda's wife, Vanessa.

Similarly to Mark Twain's ritual of reading his early drafts aloud to his wife and daughters, Miranda says that Vanessa's at-times blunt critiques were key to some of the revisions that make the final version work. And the final version did work, by nearly any standard, earning universal critical acclaim, 11 Tony Awards (including Best Musical), and a seemingly endless run of sold-out shows.

Miranda emphatically attributes his success not to genius, but to hard work. "I'm not a ****ing genius," he told Rolling Stone in 2015. "I work my *** off. Hamilton could have written what I wrote in about three weeks. That's genius."[321] Miranda, meanwhile, took seven years of rough drafts and top-down editing to bring *Hamilton* to life.

Interestingly, at least one person would disagree with Miranda's assessment of Hamilton's genius: Hamilton himself.

> "Men give me credit for some genius," Hamilton is reported to have said. "All the genius I have lies in this: when I have a subject in hand, I study it profoundly. Day and night it is before me. My mind becomes pervaded with it. Then the effort that I

have made is what people are pleased to call the fruit of genius. It is the fruit of labor and thought."[322]

Analogous Thinking

You may not be writing your own Broadway show about a founding father, but if you're looking to unearth some creative recombinations of your own, you might consider the concept of analogous thinking.

This approach suggests that often you can reach the best solution to one problem by looking at what already exists in a different domain. By adapting elements of those works to your own purposes, you can create something new, exciting, and effective.

Researcher Gary A. Davis estimates that "perhaps 80% of creative ideas"[323] follow this formation. The key is those clever recombinations. This is also why, in a brainstorming session, an outsider with limited domain knowledge about the subject but with strong expertise about some analogous field can help spur some startling discoveries.

Miranda is not the first to find grist for musicals in unlikely places; as creativity expert Davis points out, the Broadway juggernaut *Cats* was inspired by T.S. Eliot's *Book of Practical Cats*. But Miranda may be the first to reimagine George Washington's cabinet meetings as vitriolic rap battles between Alexander Hamilton and Thomas Jefferson.

In the 90's, engineer and bird-watcher Eiji Nakatsu of Japan was working for the rail company JR-west when he observed a kingfisher diving for fish while barely disturbing the water. Realizing that the head and beak shape resulted in incredible aerodynamics, Nakatsu modeled the front train car after the bird to create a quieter, faster bullet train.[324]

The shape of Pringles potato chips was inspired by how tightly and compactly wet leaves can stack on each other. (If you prefer a heartier, more old-fashioned chip, you may be able to find other parallels between Pringles and dead leaves.)[325]

The ubiquitous fastening material Velcro, beloved by parents of toddlers everywhere, came out of a fateful 1941 trip in the Alps, when

Swiss engineer George De Mestral and his dog bumped into burdock thistles. Later, De Mestral was surprised by the thistles' sticking power to his clothes and dog. Intrigued, he grabbed an old microscope and magnified a sample. The burr was covered in tiny hooks that stuck to the natural loops created by fabric or fur.

De Mestral says he instantly recognized the analogy between thistles and clothing fasteners. He's quoted as recounting his reaction at the time like so:

> "I will design a unique, two-sided fastener, one side with stiff hooks like burrs and the other side with soft loops like the fabric of my pants. I will call my invention 'Velcro,' a combination of the word 'velour' and 'crochet.' It will rival the zipper in its ability to fasten."[326]

Whether this was his exact thought process, or, more likely, how he remembered it later (keeping in mind that memory is not a perfect recorder of events), the key point is that De Mestral's analogous brain made the link.

As all good invention stories go, his bold challenge to the tried and true zipper initially brought its share of ridicule. De Mestral, however, was undaunted, and spent years employing top-down problem-solving, working through a variety of material applications, until, through trial and error, he hit upon nylon sewn under infrared light as the perfect hooks for his artificial burrs. He patented the idea in 1951 and never looked back as Velcro went on to become a multimillion-dollar business.

The Velcro idea has continued to morph across other domains, from healthcare to children's shoes to space travel. NASA's astronauts use a Velcro like product to keep their dinner plates from floating away during weightlessness in orbit.[327]

While Velcro's unexpected success is inspiring, analogous thinking has a much bigger story to tell from a pivotal year way back in 1859, when Charles Darwin published *The Origin of Species*. He arrived at his central notion of evolution by employing analogous thinking when he reverse-engineered the human practice of selective cattle breeding to better understand how a species evolves over time.[328]

Analogous Thinking in Business

In 2008, Oliver Gassmann and Marco Zeschky chronicled their research of analogous thinking in the world of product innovation.[329] The researchers were spurred on by BMW's 2001 iDrive, which used a standard video game joystick to eliminate up to 200 different knobs and switches, creating a simpler dashboard in BMW's luxury cars.

Seeing this crossover problem solving led Gassmann and Zeschky to wonder how often engineering teams were employing this technique. They decided to study companies who were using analogous thinking to spark creative breakthroughs.

One of those case studies was the AlpineCo, a prominent ski manufacturer. The company discovered that at high speed, their snow skis developed vibrations, called 'resonance frequency.' At 1800 hertz, the vibration became so intense that it made their skis extremely difficult to control, and, therefore, dangerous.

AlpineCo's Research and Development team thought they could save time if they looked to other industries that had already solved vibration problems. Without a solid idea how to refine their initial search, the R&D team simply began by kicking around general concepts on the Internet like, 'vibration, damping, and cushioning.' From the very start, they were open to and "actively looking for analogous solution,"[330] but this by itself wasn't enough.

The problem was that without any way to define a task set, the scope of the inquiry became overwhelming. Chasing down the plethora of industries that had at one time or another dealt with vibration problems proved to be untenable. Finally, one of the R&D team members realized that they might have more luck if they only focused on vibration solutions developed to counter a frequency of 1800 hertz or higher.

This led them to a rather surprising solution. It seems that this high hertz vibration problem was well known in the field of acoustics, particularly to the makers of bowed instruments. Drawing a bow quickly across strings can produce a chattering effect, similar to how it feels when you slam on your car's anti-lock brakes. In a violin, it results in uneven pitch—not quite the deadly danger of uncontrollable skis, but still a drawback that demanded solutions.

One enterprising luthier had solved the chattering problem by incorporating a deadening layer of material in his bows, which smoothed out bowing by absorbing the vibration and knocking out the resonance feedback. The R&D team at AlpineCo tested the idea and found that this same material could easily be incorporated into their skis at a minimal cost and effort. This ski dampening has the impressive name of 'frequency tuning,' and is now standard operating practice throughout the ski industry.

At first blush, it might not seem like a luthier would have much to teach a ski maker. But analogous thinking shows us that solutions in one domain can be repurposed again and again for other domains. That's why it might not be surprising to learn that in an analysis of patents, "most inventions were based on a rather small number of generally applicable principal solutions."[331]

Researchers Oliver Gassmann and Marco Zeschky say there are three important points to keep in mind when applying analogous thinking to innovation and problem solving.

First, it's critical to break down the problem and its context until you thoroughly understand exactly what you're dealing with. As in Kaizen, an in-depth analysis is key to determine the root cause. In the case of the ski company, this meant they needed to understand just what was happening to their skis at high speed. This, in turn, leads to abstractions from the problem into specificities like vibration, dampening and cushioning.

Second, you can't access solutions across domains unless you're willing to think outside the confines of your immediate subject. Even though the AlpineCo R&D team did not have exposure to the field of acoustics, the R&D leader opened the door for looking at other possibly relevant opportunities. This deliberate break with conventional thinking is necessary to allow for the identification of novel solutions. The Greeks called it 'synectics,' "the joining together of different and apparently irrelevant elements."[332] This is the real power in analogous thinking.

Finally, you have to carefully assess your results to ensure that the repurposing of the idea has merit. AlpineCo ran material tests to satisfy themselves that indeed there was a crossover between instrument bow dampening and improved ski design.

Gary A. Davis suggests that to better employ analogous thinking in your problem-solving, you should ask yourself these four important questions:[333]

> What else is like this?
>
> What have others done?
>
> Where can I find an idea?
>
> What idea can I modify to fit my problem?

If you frame your world in this way, you open yourself up to exploring more domains. By comparing and contrasting new information to your own domain expertise, you can hit upon some uniquely creative ideas.[334]

In your culinary adventures, you might not think about mixing chocolate with bacon, but many people swear by it. Paul Simon's *Graceland* melds traditional American roots rock with Cajun zydeco and South African mbaqanga, among other influences.[335] Music and fashion icon David Bowie's inspirations included sci-fi novels, miming, the Japanese theater tradition of Kabuki and avant garde performance artists, to name a few.[336]

Eli Whitney, of cotton gin fame, was the first to figure out how to clean short-staple cotton, and he drew his inspiration from watching a cat trapped on one side of a wire fence trying to pull a chicken through the fence.[337] (A pretty rough day for the chicken, but a great day for Whitney.)

Medici Effect

If you're a fan of the Renaissance, you are probably familiar with the House of Medici.[338] This powerful political dynasty bankrolled generations of thinkers, poets, philosophers, sculptors, painters, architects, and scientists. It is no exaggeration to say that in 15th century Italy, the Medicis were a driving force behind making Florence, well, Florence. We still enjoy their creative legacy; Medici sponsorships enabled the work of heavy hitters like Galileo and Botticelli, as well as Donatello, Raphael, Michelangelo and Da Vinci. (The artists, not the Teenage Mutant Ninja Turtles.)

The innovations of the Renaissance, Frans Johansson argues in his book *The Medici Effect*,[339] were partly spurred by the natural result of funding and gathering so many intellectual and artistic masters in one place, where they rubbed shoulders and exchanged ideas. According to Johansson, the Medici genius was in creating conditions that fostered the intersection of diverse disciplines, enabling analogous thinking and thus ratcheting up Kirby Ferguson's copy, transform and combine process. This co-mingling ultimately led to extraordinary creative leaps in innovation.

These intersections have occurred throughout time—more recently, in California's Silicon Valley. As more and more people moved to take advantage of the dot COM explosion, a culture began to emerge that fostered a new way to think about the possibilities of computer technology. And that can foster a collective momentum, a kind of group growth mindset. Experts teach each other, challenge each other, and inspire each other. When people share passion and proximity, creativity becomes contagious.

Often when we think of creativity, we might think of it on an individual basis: the lone wolf like Velcro's De Mestral, relentlessly pursuing his passionate idea, while others ridicule and deride. Let's face it, there is something intoxicating about the legend of the underdog, and the "I told you so" moment when perseverance triumphs over the doubters.

Still, when those individuals converge on epicenters like Florence and Silicon Valley, creativity multiplies by force. And although it is true that individuals have created some incredible innovations, a disproportionate amount of creative breakthroughs occur in clusters around creative centers, where analogous thinking and the trading of domain knowledge becomes the common currency.

Johansson suggests you can create your own little Medici effect by correctly setting up your environment. The real power lies in the brain's ability to combine domain information from all walks of life. It's at these nexus points that you can spark some surprising insights. One of Johansson's central themes is to immerse yourself in as many different experiences as possible, with an eye toward seemingly unrelated connections to create new insights.

Identifying these points of connection is at the heart of analogous thinking and Kirby Ferguson's creative model of "copy, transform and combine." Johansson suggests that the key to novel ideas is keeping an aggressively open mind in the recombination phase, what he calls "the intersection", bringing together elements that might, on the surface, seem out of place. (For instance, Alexander Hamilton and Jay-Z.)

As Johansson says, "The world is connected and there is a place where those connections are made—a place called the Intersection. All we have to do is find it... and dare to step in."[340]

Superlinear Scaling

One of the intriguing things about large cities is that they produce a disproportionate number of patents and inventions. Theoretical physicist Geoffrey West demonstrated that large cities follow a power law known as 'superlinear scaling': they generate more creativity, and at a faster rate.

One might imagine that a large city means more people, and more people would, by definition, mean more creativity. But West's research showed "the average resident of a metropolis with a population of five million people was almost three times more creative than the average resident of a town of a hundred thousand."[341]

So what's going on? Researchers from the MIT Media Laboratory's Human Dynamics Lab think they know the answer.[342] Their work indicates that greater population density means more face-to-face interaction. And face-to-face interaction means greater chance of ideas becoming contagious. It's the same principle that drives the common cold, and the Internet. The greater the network, the faster and more forcefully ideas or germs can spread.

When novel thinkers congregate in one area, they have the possibility of trading, spreading, and acquiring new ideas every time they make a morning coffee run. They could encounter a colleague. They could encounter a rival. They could encounter a friend of a friend, working in a somewhat adjacent field, whose observations could spark an entirely new train of thought.

Steven Johnson says to think about your brain as a 'network' of roughly 86 billion neurons. "By comparison there are somewhere on the order of 40 billion plus pages on the Web—that means you and I are walking around with a high-density network in our skulls that is orders of magnitude larger than the entirety of the Web."[343]

When we talk about creativity in the abstract, it can sound almost too simplistic– that everything could come down to something as simple as conversations. But when you consider how often innovation lies at the surprising intersection of two seemingly disparate ideas, you begin to understand the potential power of face-to-face interaction for cultivating creativity.

Cultivating Creativity

When it comes to cultivating creativity, Scott Barry Kaufman and Carolyn Gregoire say that beat writer Jack Kerouac had it right: "The best teacher is experience."[344] According to them, the single most consistent trait that predicts creative triumph is "the drive for cognitive exploration of one's inner and outer worlds."

We've talked about this in terms of the rage to master, as well as Dweck's model for growth mindset, but Kaufman and Gregoire take a more nuanced look at growth mindset. They discuss the idea of 'openness' to learning in three distinct areas: intellectual engagement (problem solving), affective engagement (connecting to a subject emotionally) and aesthetic engagement (the seeking of beauty in the arts).

We can think of intellectual engagement as correlating to scientific creativity, while affective and aesthetic engagement are more connected to artistic creativity. Kaufman and Gregoire's research suggests that when it comes to producing creative thinking, openness trumps both IQ and divergent thinking.

At a neurological level, openness is driven by neurotransmitters like dopamine. Kaufman and Gregoire say that simply characterizing dopamine as the "feel-good" hormone is selling this versatile neurotransmitter short. Dopamine also seems to facilitate plasticity in thinking, which drives greater associative algorithms.[345]

Kaufman and Gregoire suggest that, beyond the notion of reward, dopamine promotes a *wanting*, a kind of unremitting need, the kind of desire associated with addiction and the rage to master. In this regard, dopamine is less about a physically pleasurable payoff and more about what psychologist Colin DeYoung of the University of Minnesota calls the satisfaction that comes with exploration and mastery. She labels dopamine as, "the neuromodulator of exploration."

As discussed, identifying patterns to predict future outcomes is more than a hallmark of associative memory; it's an evolutionary advantage. Recognizing the connection of the changing seasons and food availability allowed our ancient ancestors to develop migratory patterns during the cold season, a pretty important chapter in the 'stay alive' hand book. Here, dopamine is clearly much more than a feel-good drug.

We learned in Chapter Four that focus is an essential ingredient in building a solid practice app, but Harvard psychologist Shelly Carson and her team's research led to an unusual discovery about dopamine's effect on focus and creativity.

Reduced Latent Inhibition

Reduced latent inhibition can be described as difficulty filtering out interruptions in everyday experience, or the inability to stay focused. Your senses are constantly under assault: sirens, an office worker chewing loudly in the cubicle next to you, the smell of cologne in a crowded elevator, the roar of a crowd, or the buzz of email. Distraction is ever present. Obviously, these annoyances diminish our ability to focus, which in turn, might negatively affect our ability to generate creative ideas.

It's true that diminished focus might be a negative net result of heightened sensory awareness, but Carson's team discovered something unusual when they looked at Havard staff members well known for creative prowess. It seems that "The university's eminent creative achievers were seven times more likely to have reduced latent inhibition—meaning that they had a harder time filtering out seemingly irrelevant information and continued to notice familiar things."[346]

So how could a propensity for distraction lead to greater creative output?

Rather than a detriment, researcher Darya Zabelina of Northwestern University says that having a 'leaky' sensory filter—a brain that struggles to filter irrelevant information from the environment—is linked with higher levels of creativity, because these people simply process more information than their better-focused counterparts.[347] And the more information a person can process, the wider the set of associative connections, which leads to a greater possibility of novel solutions.

History is rife with examples of well-known creatives and their easy distractibility.[348] Charles Darwin, Marcel Proust and Franz Kafka all are reported to have had a hypersensitivity to sound; something as simple as a loud ticking clock could derail their thinking. But the flip side of their heightened sensory awareness was that they were less likely to miss out on the linkage of everyday experience and its connection to more powerful ideas.[349]

Problem Finders

In the 1960's, social scientists Jacob Getzels and Mihaly Csikszentmihalyi were searching for the root cause of creativity. Their quest brought them to the Art Institute of Chicago, where they observed fourth-year students preparing to draw a still life by arranging standard drawing class objects on a table. In *To Sell is Human,* Daniel Pink describes what happened next:

> "The young artists approached their task in two distinct ways. Some examined relatively few objects, outlined their idea swiftly, and moved quickly to draw their still life. Others took their time. They handled more objects, turned them this way and that, rearranged them several times, and needed much longer to complete their drawing. As Csikszentmihalyi saw it, the first group was trying to solve the problem: How can I produce a good drawing?"(solely about the technique) "The second was trying to find a problem: What good drawing can I produce?" (about creating art)

In a subsequent art show, a panel of experts declared that the problem finders generally had the better drawings. Follow-up studies eighteen years later confirmed the finders were indeed more successful in the art world.[350]

According to Pink, "Getzel and Csikszentmihalyi's research influenced the modern understanding and academic study of creativity. In subsequent research, they and other scholars found that people most disposed to creative breakthroughs in art, science, or any other endeavor tend to be problem finders." These people tend to experiment more across a variety of disciplines, search for unique combinations, and show flexibility in both their approach and their willingness to change course as necessary.

Creativity in Motion

If you were going to try to imagine a problem finder's lair, the hideout of a person who created a piece of technology that revolutionized the fitness world and left a worldwide cultural impact, you probably wouldn't picture a barn in rural Minnesota.

But to be fair, you've probably never driven out to the wide-open farmlands of small-town Waconia, past an electronically operated gate and down a long dirt road to visit former hockey goalie and full-time entrepreneur Scottie Olson.

Chickens roam the grounds. Swans bob around in the lake out back, their honks creating a sort of free-form jazz. But the focal point of the yard is definitely the large loop of elevated track, like a misplaced section of the Chicago 'L'. It's a full-size prototype of a contraption somewhere between a ski lift and a gentle roller coaster. *SkyRide*, a sign proclaims.

After that, the white barn perched atop a hill, with its swooping green metal roof, seems relatively normal. Except, perhaps, for the oversized replica of a trophy-mounted zebra head above the double doors.

"When I bought [the barn] it had piles of hay in it," remembers Olson. But he'd known right away it would be one of his future homes. "That was always my dream," he told us, "to live in a barn."

Olson is a man who trusts his instincts, whether he's developing the SkyRide or commissioning his artist friend Jimmy Hartman to design a penguin lawn ornament—or creating the Rollerblade. If you've ever strapped on a pair of Rollerblades and gone racing through your neighborhood, you have Scottie Olson to thank.

The idea of attaching a row of wheels to a shoe is not a new one. In-line skates have existed since at least the mid-1800s. In 1849, when German composer Giacomo Meyerbeer wanted characters in his opera *Le prophète* to appear to ice skate onstage, he had a butcher named Louis Lagrange whip up a primitive version of the wheeled skate, although this model had some essential drawbacks. For instance, it was almost impossible to turn or stop, two fairly important features in any skate design.[351]

CHAPTER 6

We tend to think of the clunkier looking quad skates (also known as "roller skates") as a primordial ancestor to the sleeker in-line skate, but in fact, at the time, the quad skates' double row of wheels was an innovation, since putting more weight on one side or the other allowed 19th century skaters something like steering. For the next century, quad skates left their in-line forbearers in the dust.

In-line skates survived here or there as a curiosity, but they were unknown to mainstream society when Olson tried on his first pair as a teenager in the late 1970's. Olson was back in the U.S. after playing as the only American in Canada's Junior A ice hockey league, even making it to the Memorial Cup. "I had a really good career as a goalie," he remembers. "And then I signed a contract and played in the farm league down here, in the States, but I always had that idea that I was gonna be my own boss doing something. And then I happened to fall into the blades."

It was love at first skate for Olson, who immediately saw the potential. Here was a way to free ice skating from the ice—to skate anywhere and in any weather. "And then I realized that, as great as they were, they were sure slow," he says. In fact, early in-line skates were slower than roller skates. "They were made by a guy out in LA, who had never skated on them," he explains. Olson, on the other hand, brought with him the speed demon expectations of a serious ice hockey player. He figured there had to be a way to make them go faster.

Olson got to work. He and his brother began to tinker with the initial set-up of the skate, remaking it with modern materials. They added a rubber toe break and polyurethane wheels, and after some experimentation, borrowed their ball bearing design from roller skates. In a stunning victory for analogous problem solving, they'd found that the roller skate's ball bearing configuration was also the most effective one for this new type of skate.

Inspirations came from other athletic gear as well. "One of the big reasons the Rollerblade became a big recreational skate was the type of boot we ended up using," he says. "We got lucky, because just at that time, the first plastic boots were being produced [for ice hockey]." This itself had been a product of analogous thinking: "They took the technology from ski boots, which used to be leather."

And so Olson had a Medici moment of sorts, traveling to Italy, which was then at the forefront of the plastic athletic boot scene. Olson was thus able to borrow the plastic boot design from ice skates, which had borrowed it from skiing. "That way, a recreational person could get that support you really needed."

Olson describes his work with Rollerblades as a combination of top-down and bottom-up thinking. Like a younger, more physically fit Edison, he says the process included a number of promising ideas which didn't go anywhere. Still, he doesn't see himself as the inventor of the in-line skate; after all, other models had already existed. But by 1981, the new innovations, including those clever recombinations of features from other forms of skates, had created something all its own: a Kirby Ferguson-style remix you could wear on your feet.

Olson was equally ingenious in marketing. His ice hockey friends used Rollerblades for off-season training. He organized roller hockey teams. And he spread the word by skating everywhere he went—sidewalks, cities, beaches, even airports. In the days before the TSA, he said it wasn't uncommon for him to Rollerblade right down the aisle of an airplane. People took notice.

The timing was right for it, he adds, unknowingly reflecting Daniel Kahneman's truism about hard work plus luck. Rollerblades provided an alternative to ice skates for American athletes tired of the long-running Canadian monopoly on ice hockey. And they were a sleeker, cooler alternative to roller skates for a public that was just beginning to go crazy for outdoor fitness.

By the early nineties, Rollerblades was a $650 million sensation—and Olson was no longer with the company. The friend he'd hired as his accountant had been embezzling, and they were far behind on their taxes. The resulting financial straits led Olson to turn to some wealthy local businessmen for help, who ultimately bought him out of his own company, in a move that was technically legal but which arguably played on the business naiveté of a young guy from Minnesota.

However, Olson left Rollerblade as a multi-millionaire, and a man determined to keep creating. He credits some of his determination to his days as a goalie in Canada—the lone American defending the goal, out on the ice by himself.

> "You have to try not to let anything really get you down," he says. "And pressure doesn't bother me. I learned, somehow, to deal with it. I do know exercise certainly helps."

Olson's newer inventions include the RowBike, which marries the concept of a rowing machine with the mobility of a bike to create a form of transport that provides an upper body workout on two wheels. It's beloved by men over fifty, he notes, as well as the disabled community. "That's the beauty of rowing," he says. "You can be a paraplegic and still row."

His current project is the SkyRide, the track-mounted bicycle-powered device that allows riders to pedal in the air. "I don't just want to make fitness fun," he says. "I want to make it exciting." When we caught up with Olson in late 2015, he had just sold his invention to the Carnival Cruise line. "We're actually building, right now, the track that is going up on the ship," he explains. "And we're building the rides here, in our shop. So we're on the fast track now, because all of a sudden, it's gotta get done. Enough dinking around with prototypes. We've gotta build production models."

Talking with Olson, one gets the impression of a man who is always dreaming about his next big idea. Conversation is filled with long pauses. Olson certainly seemed to be a model of reduced latent inhibition, interrupting the interview several times to point out a group of young swans, to admire a hawk in flight, to let a pack of pet dogs in and then out of the room again, and then to scoop up one large dog under its front legs and lift it into his lap while he spoke, as if bouncing a giant baby.

If it made Olson something of a unique interview, these tendencies may also explain some of his success. Perhaps his freewheeling approach to attention allows him to make observations the rest of us miss, watching the birds wheel overhead and allowing the sight to capture his imagination with a new, surprising thought experiment about kinetic possibilities. Olson himself noted that his best ideas tend to come to him while he's in motion. "I work out in the morning, and I always have my phone or my pen and paper, because it gets me thinking better."

Legions of rollerbladers, disabled athletes, cruise line goers, and penguin lawn ornament fans thank him for it.

Chapter Key Points: Hacking Your Creativity App

- The first step of creative thinking is to develop a deep level of domain knowledge.

- Leverage your own Medici effect. By combining separate spheres of domain knowledge, you create a situation where the whole is greater than the sum of the parts.

- One way to spur analogous thinking is by employing Gary A. Davis's essential questions:

    ◦ What else is like this?

    ◦ What have others done?

    ◦ Where can I find an idea?

    ◦ What idea can I modify to fit my problem?[352]

- Kirby Ferguson's copy, transform and combine model serves as a template for creative behavior.

- Creativity requires a top-down thinking approach, which often results in some degree of failure.

- Remember, epiphanies are not the result of creative lightning, but are the direct consequence of a stymied top-down process.

- For best results, cultivate the flexibility, experimentation, and novel attitudes of a "problem finder."

Chapter 7

Flow App

It's been a long journey, from mindset to creativity. Having detailed much of the steps for success and explored the necessary brain apps, now it's time to head for the finish line, the final payoff for all the hard work: flow.

In sports, it's called 'being in the zone'. Musicians sometimes describe it as 'being in the groove.' It's those moments when self vanishes, and you're operating with maximum confidence. A sense of calm pervades, even though you might be surrounded by a frenzy of activity. Your focus intensifies and your actions and decisions seem to meld. This is the psychological state known as flow, researched and popularized by psychologist Mihály Csikszentmihalyi.

You might know what flow looks like from the outside if you've ever experienced a great jazz player riffing, a wide receiver leaping to snag the football with his fingertips, or a skilled comedian doing killer improvisation. A writer experiencing flow has the sense that some outside source is dictating the words on the page. Runners, skydivers, surfers, programmers and students engaged in deep academic pursuit are all secret, or not-so-secret, flow junkies.

Csikszentmihalyi says that flow state involves a degree of intense concentration, but it's also so much more: "The ego falls away. Time flies. Every action, movement, and thought follows inevitably from the previous one, like playing jazz. Your whole being is involved, and you're using your skills to the utmost."[409]

Flow separates the high points of experience from the mundane. In this state of singular focus, nothing seems to matter but your present activity.

So just what's happening inside the brain during a flow state? In 2008, Johns Hopkins neuroscientist Dr. Charles Limb[353] set out to answer that question. He began doing experiments with jazz pianists. Limb had them improvise music on a specially constructed keyboard while lying in an fMRI. He had hoped that, by measuring blood flow, he could tease out which neural mechanisms were activated during an improv session.

Limb observed that when a jazz player got into a groove, blood flow seemed to decrease in an area called the dorsolateral prefrontal cortex, or dorsolateral PFC.

You may remember the dorsolateral PFC is the home of executive control, where your inner critic resides, the locus of top down thinking. This area helps to control your impulses, monitor behavior, and analyze your actions. It tracks meta data, reviews choices, and compares and contrasts, all in an effort to make logical deliberate decisions. Lowered activity in the dorsolateral PFC would seem to suggest that your inner critic has gone on vacation.

Is there a word to describe the ramping down of the dorsolateral PFC when cognition takes a backseat to practiced behavior? Yes: hypofrontality. Another way to put it: a loss of your sense of self. It's when the jazz player becomes one with that piano or horn, a single entity from which the music seems to spring forth.

Where does all that decreased blood flow go? Limb observed increased movement in the medial PFC, home to self-expression. This part of the brain fosters a sort of ungoverned creativity, where rules are more like general guidelines, and impulse and action rule supreme.

Another area of the brain that helps keep tabs on your behavior is the superior frontal gyrus, or SFG. It's associated with self-regulation, and the inhibitions that keep you from acting in ways that might get you in trouble. Essentially, your 'good angel' doesn't sit on your shoulder like in the cartoons; it lives in your brain's SFG. In *The Rise of Superman: Decoding the Science of Ultimate Human Performance*, author Steven Kotler cites an Israeli study that says when people are deeply immersed in a task, from having sex to playing cards to climbing a mountain, their superior frontal gyrus also begins to deactivate.[354]

So in theory, Kotler says, as the dorsolateral PFC and SFG quiet down and the medial PFC ramps up, there is a trade-off: less energy for analytical cognition and more energy for concentration and focus. Concentrated practice and focus is at the heart of flow state. Without executive top-down scrutiny, action becomes spontaneous and in the moment.

This is the essential definition of the flow state, in which a cocktail of neurochemicals rush in and take charge. Norepinephrine increases emotional control and focus. Dopamine generates an increase in blood flow, pattern recognition and neural efficiency. Dopamine also drives positive feelings toward goal fulfillment. Endorphins kick in to minimize pain and promote a euphoric feeling, allowing you to pursue a goal even though there might be extreme mental or physical discomfort involved. Anandamide reduces fear and increases lateral thinking, what we referred to in Chapter Six as analogous thinking.

Anandamide may be the least known of these chemicals, but it's certainly not the least important. It promotes rapid associations, drawing on the domain knowledge stored in your experiential memory, which in turn allows your brain to create unusual connections and surprising solutions. It's a basic tenet of any act of improvisation, or at least, any interesting improv.

The final cherry on the focus sundae is the neurotransmitter serotonin, enhancer of the warm afterglow feeling. Although it's not released during flow state, it's part of the feel-good reward system that helps habituate the flow behavior for future repetition.

Sometimes we hear people, especially in extreme sports, described as 'adrenaline junkies,' but as you see, it's a little more complicated. The end result of this chemical piling-on is a state of altered consciousness. It's taken western science and the fMRI to finally catch up, measure, and confirm what Zen masters and jazz players have experientially understood for years.

People who enter the zone or find themselves in flow state say they suspend thinking. That is to say, they are simply experiencing the event as it unfolds. They rely on instinct, using processes driven by the brain's reflexive System 1. In this mode, your carefully rehearsed neural code is playing unconsciously. When in flow, it feels like you're simply along for the ride.

No deliberate analysis is needed, or even desired. When Dave Guy is improvising jazz, there is no way for him to be both in flow and at the same time second-guessing the next note he's about to play. Like the U.S. men's 2012 gymnastic team in Chapter Four, trying to observe yourself in the middle of performance knocks you out of flow.

The good news is that entering a flow state doesn't require a pommel horse or full mastery of the trumpet. If you are drawing a picture, and you're approaching the boundaries of your skillset, concentrating on each and every stroke of the pen can shift you into flow. That's true for any domain that you seek to master. Any motor/sensory activity that pushes you to the edge of ability and focus can move you into flow, from playing chess to writing computer code—even raking leaves or shoveling snow.

In his book, *Flow: The Psychology of Optimal Experience*,[355] Csikszentmihalyi suggests that the key to living a life of fulfillment is to get in the habit of challenging yourself in ways that will more often bring you into this state.

Flow expert Steven Kotler says that if you want to hack flow, you need to pay attention to the triggers. As we learned in Chapter Five from B.J. Fogg's work on habit, triggers are the all-important start buttons for behavior initiation.

Psychological Triggers

When seeking flow, the first step is to have a clear goal in mind. Following the **SMARTR** system from Chapter Two can provide a useful framework. Take the time to figure out what exactly you want to accomplish. This means both setting your long-term goal, and designing a deliberate practice program that will allow you to make all the gradual improvements you'll need to get there.

Flow follows the Goldilocks rule: not enough difficulty, and you'll get bored. Too much, and you'll quit out of frustration. You need to seek a middle ground, where the action is difficult, but still possible. If I challenge a friend to a game of chess and I beat him handily, I might initially enjoy the experience, but over time, it's likely I'll lose interest, and so will my friend. On the other hand, if I challenged Grandmaster Susan Polgar to a chess match, I'd be annihilated right away and the game would be a bust for both of us.

There is a sweet spot where your skill level is appropriate, the challenge is real, the rules clear, and the outcome uncertain. "Enjoyment appears at the boundary between boredom and anxiety, when the challenges are just balanced with the person's ability to act,"[356] writes Csikszentmihalyi. This is the essence of pursuing a flow experience. You just need to build a brain app that pushes you toward the next level without destroying yourself.

Flow Requires Hypofrontality

Imagine a bird flying through a turbulent air pocket. Although we can't know for sure what's happening in the bird's brain, we can assume that the bird isn't actually analyzing thermals, aerodynamics, or each individual flap of its wings. Rather, we might posit that thanks to a lifetime of flight experience, the bird can use the correct wing motions without stopping to carefully weigh the pros and cons. The changes come intuitively.

In essence, a lifetime of practice allows that bird to tap into stored neural mapping and execute whatever adjustment it needs to make effortlessly—the very definition of hypofrontality.

When a young Susan Polgar played chess against other established chess champions, she was testing her limits and tapping into a rarified level of challenge. And it's this kind of competition that forced Susan to push herself into unchartered waters and test her perceived boundaries of ability. Each move and countermove was providing real-time feedback, creating a powerful opportunity to fine-tune her behavior moment by moment. That's a very different process from analytical deliberation.

For humans, those preloaded bits of domain knowledge inform second by second adjustments from chess games, to jazz improvisation. This is why it's critical to take the time to load those valuable strings of flawless neural code through deliberate practice.

In those competitive moments where the boundaries are pushed and then exceeded, the Bannister Effect takes hold. In the aftermath, you realize that you are capable of operating at a new higher level of performance. This is not only the hallmark of growth mindset, but also the manifestation of flow.

New challenging situations train you to enter flow more often. Dave Guy's decision early in his career to stretch his ability and play as many musical genres as possible helped set his brain up for future flow. Susan Polgar's play against a wide variety of chess competitors had the same consequence. All of this improves your associative creativity, thus compounding your chances of experiencing even more flow.

Complacency doesn't lead to flow. This is why you can experience flow while stretching your skillset, but once you've mastered that new skill, the same situation will no longer provide the same feeling. For this reason, athletes who use extreme sports as their gateway to flow are always pushing their limits and sometimes, unfortunately, increasing the danger to life and limb.

The Merging of Action and Awareness

When you are in flow and bringing all your energy to act at your highest skill level, you become completely unaware of everything else around you. In a flow state, there is a loss of self, and of self-judgment. There is no bandwidth for the inner monologue that foments doubt or arbitrates our behavior, whether good, bad, or indifferent. As Csikszentmihalyi says, "There is no excess psychic energy left over to process any information but what the activity offers."[357]

When action and awareness merge, you experience what's known as 'deep embodiment': all your senses come alive. And as our senses are activated, the experience broadens and consumes us.

We generally think about our senses in terms of sight, hearing, taste, smell, and touch, but two other senses come into play during flow: proprioception (your sense of your body's position in space) and vestibular awareness (your sense of balance). If you've ever experienced vertigo, then you know how it feels to momentarily lose your proprioception and vestibular awareness. This occurs when your vision and balance are no longer operating in sync. The result can be extremely unsettling—for instance, creating the sensation of being trapped in a spinning room.

Flow, however, seems to heighten proprioception and vestibular awareness, giving the sense of total command over the body. Flow practitioners say it's impossible to describe the sensation, but that it's highly addictive.

After a flow experience, you might have the chance to go back and replay the event in your mind. But often, people who have experienced flow will say that they don't actually remember it moment by moment.

Group Flow

Flow is not restricted to an individual. When two or more people experience flow together, this state is known as 'group flow.' In Chapter Six's discussion of Renaissance artists, the Medici Effect likely fostered a certain amount of group flow.

When a sports team enters the zone, momentum shifts. This is probably the most common way we think about group flow, as our favorite team makes an improbable comeback to snatch victory from the jaws of defeat, and each player seems to move in perfect choreography. This unification generally occurs when the players share a profound understanding of their domain, and of each other. In this setting, ego tends to fall away, as does single-player domination.

This is often the product of good coaching. During the 90's when the Chicago Bulls were making their run for NBA championships, some suggested that coach Phil Jackson's primary task was to convince superstar Michael Jordan that he couldn't win the game by himself.[358] Jordan's rage to master had taken him far, but Jackson helped him understand that he still needed to pass the ball. As Jordan's game matured and he learned to trust his team, the Bulls went on to win six NBA titles.

But it's not just team sports. Kotler points out that group flow is very common in the business world—the successful businesses, at least.

Salim Ismail, former head of innovation at Yahoo, back in the days when Yahoo was still an innovator, said:

> "Because entrepreneurship is about the non-stop navigation of uncertainty, being in flow is a critical aspect of success. Flow states allow an entrepreneur to stay open and alert to possibilities, which could exist in any partnership, product insight, or customer interaction. The more flow created by a startup team, the higher the chance of success. In fact, if your startup team is not in a near-constant group flow state, you will not succeed. Peripheral vision gets lost and insights don't follow."[359]

To keep the flow going, a company must constantly reorient itself toward improvement. This is the lesson from Chapter One that Blackberry failed to understand, and, ironically, Yahoo eventually fumbled as well. It also speaks to the impermanence and elusiveness of flow.

Flow in business can have profound effects. Kotler cites a 10-year McKinsey study that found:

> "Top executives are five times— i.e. 500 percent— more productive in flow. In studies run by the U.S. military, snipers in flow learned between 200-500 percent faster than normal. The study says that in a flow state your creativity gets a 7x boost."[360]

Creativity As a Flow Trigger

Psychologist Arne Dietrich suggests flow is the third rail of creativity, separate and distinct from top-down and bottom-up thinking, which are both part of your analytical System 2 process.[361] The information you rely on during a flow state is free from the scrutiny of your PFC's executive control. While in flow, the neural code that's directing your action is actually embedded in the physical motor activity itself. This should not be confused with the common misconception of 'muscle memory'.

When we spoke with Dietrich, he explained:

> "The memory of how [muscles] are being contracted—is in areas of the brain called the basal ganglia and the supplementary motor cortex is involved. There are sections or part of the cerebellum that keep these memories, but because they're implicit, that means they're also unconscious. It feels as if they are in your muscles, because your body seems to know what it needs to do. But where the memory is actually kept is in the brain, and that drives muscle contraction."[362]

Dietrich says, "The essence of flow is the merging of perception and action, the smooth, rapid-fire integration of sensory input and motor output that cleanly bypasses the centers of higher thought and consciousness."[363]

Take riding a bike, for example. You learned to ride a bike through trial and error, eventually myelinating enough neural code to steer, pedal and keep yourself balanced at the same time.

Successful riding happens when your brain's wiring, through anticipation and feedback from your senses (sensory input), connects your muscles (motor output) to the bike and road conditions. And if it all works out, you end up staying upright on the bike.

The process of sensory and motor integration happens below consciousness. You might experience falling off your bike and skinning your knees a few times first, but you don't experience the 'discussion' between your sensory and motor systems as your brain fine-tunes your internal mapping for bike riding. That mapping exists below your awareness.

This is why it's virtually impossible to explain to someone how to ride a bike. The wannabe rider might be able to glean clues from verbal directions like, "pedal faster", or "keep the front wheel straight," and they might be able to learn from watching, but the bulk of information isn't available through secondary abstraction. It is only available in the context in which it was acquired, through the practice of riding. You can't learn to ride a bike by reading a manual. Or we might say that the 'riding manual' is contained in your skull, and assembled in the brain, again and again with each riding experience.

This turns out to be true about any physical activity, from dunking a basketball to tying your shoes.

Try teaching someone to tie his or her shoes with only verbal commands. It's virtually impossible. Physical mastery comes from rehearsing and practicing a particular movement. Flow happens when movement has become so automatic and so seamless for the practitioner, that they fully embody the activity in the moment of action.

Dietrich explained musical improvisation and flow. Once you master the ability to play each note, you can then suspend analytical thought and recombine the notes in an interesting unique pattern. He told us:

> "The motor output from the sensory input, the flow experience, has to be completely automated. Only then can you produce the sequence. And the sequence is actually the creative part, not the actual output of each and every step of the motor sequence. The only way you can be creative in the flow experience is that the motor sequence produces a novel pattern.
>
> But each step of that pattern has to be automated. Otherwise it cannot be in flow. If you have to engage your explicit system [analytical System 2] your flow is gone. Because obviously, that's the whole point—that your consciousness is disengaged. You have to have a state of hypofrontality, where your explicit system does not interfere."[364]

One might expect that in flow state, the brain must exhaust its resources to improvise jazz. But, as Dietrich suggests, that's not the case. In fact, it's just the opposite: the brain does not consume large quantities of glucose to fuel the flow state, but instead operates more on low energy cruise control, denoted by hypofrontality.

Environmental Triggers

Environmental triggers can also lead to flow experience. Our brains are wired for survival, and as a result, when we are put in situations that we deem dangerous or unsafe, our amygdala's fear center goes into overdrive. In a heightened state of fear, we are far more likely to operate out of our reflexive System 1. But even if we've never actually experienced that exact situation before, lateral thinking allows us to escape a variety of life-threatening situations by tapping the flow state.

Flow in Aerobatics

Aerobatics is just what it sounds like: out-of-the-ordinary, acrobatic flying maneuvers performed in an airplane or helicopter, frequently involving rolls or spins. When executed at low altitudes for an audience, the term is "stunt flying." However, it's about far more than just wowing a crowd on a Midwestern fairground.

Aerobatic competitions are a refereed sport, flown at loftier—and thus much safer—heights, under strict rules, with judges weighing in on the quality and precision of the aerial feats. The sport owes many of its signature moves to the dogfighting days of World War One and World War Two, when air battles could be won or lost based on a pilot's ability to outmaneuver and outshoot enemy pilots.

Even now, pilots train in aerobatics to hone their skills and prepare themselves for unusual circumstances. It can be a matter of safety: most pilots will likely never need to make a very sharp U-turn thousands of feet above the ground, but should such a situation arise, they definitely don't want to be doing it for the first time. Even a competitive aerobatics champion sometimes faces those dreaded unusual circumstances.

How unusual? How about realizing the wing might be about to fall off your plane?

On June 3, 1970, a trophy-winning former RAF pilot named Neil Williams set out on a practice flight for the upcoming World Aerobatics

Championship. He was flying the highly maneuverable Zlin Z526, a durable Czech sports plane.

Late that June afternoon, when pulling out of a vertical dive, Williams heard a loud bang—never a good sign for a pilot—and felt the airframe jolt violently. "At the same instant," he would later write, "there was a sudden and very peculiar increase in the slipstream noise, and I found myself leaning against the straps to the left, although, as I looked left, the aircraft appeared to be flying level."[365]

With no time for analytical thinking, Williams began to operate reflexively instead.

The 35-year-old had developed plenty of technique; he had been flying for more than half of his life. According to a later report by the Accidents Investigation Branch of the British Department of Trade and Industry, by 1970, he had accumulated 6,500 hours in the air, and had been active on the international aerobatics competition scene since 1962.

All of that domain knowledge was about to come in handy, to say the least. The accident report would ultimately show that a left wing spar had failed, causing the wing itself to fold up to at least a 45-degree angle from the body of the craft. As Williams soon realized, the wing was flying level; he and the rest of the plane were not. Also, he was beginning to lose altitude. If that wasn't bad enough, he had no parachute.

These moments could easily flip from dangerous to catastrophic. For Williams, on that day, the pressure did something else: it released a trove of neurochemicals. As his levels of norepinephrine intensified, so did his focus. Dopamine flooded his brain, increasing his awareness, and fortunately for Williams, his anandamide levels began to rise.

Williams first reduced power and centralized control. He instinctively grabbed the throttle to reduce air speed and thus the flight load. In response, the nose dropped lower, and the plane only rolled further to the left. He attempted to correct by applying full power; nothing happened. He was losing control of the aircraft and still falling fast.

Lucky for him, this must have been about the same time anandamide finally activated in his brain. Williams' lateral thinking kicked in and he suddenly drew a connection to a report from Bulgaria he'd read years earlier: another pilot had been flying upside down in a Zlin when a top wing bolt had failed, causing a similar situation. How had it been resolved? In that case, "the aircraft had involuntary flick rolled right way up," Williams remembered, "whereupon the wing came back into position, and the aircraft was landed by a very, frightened, but alive, pilot."

Williams guessed that the inverse had happened to him—a broken lower wing bolt. "It seemed possible that if positive G had saved the Bulgarian," Williams wrote, "negative G might work for me." In other words, he decided that his best bet was to roll upside down and hope gravitational force would push the wing back into place. "In any event," he noted, "there was nothing else left to try."

And so Williams inverted the plane. The good news: the wing snapped back into proper alignment. The bad news: it snapped back with a terrible bang that suggested further structural damage. If he wanted to keep the wing attached to the plane's body, he needed to reduce air speed.

On the other hand, the nose was finally starting to rise again, just in time to avoid a collision with some trees. "I was just beginning to think I might make it after all," recounted Williams, "when the engine died."

Williams made a fast switch to reserve fuel, only to realize a second later that in this case it meant taking fuel "from the bottom of the gravity tank, which was of course now upside down." Instead, he selected the main tank once more, "and after a few coughs the engine started and ran at full power."

Now, he just needed a safe way to return to the ground. The trouble was, with no way to assess the extent of the damage, Williams worried that the strain of extending his wheels might sever the wing for good. Also, let's not forget, he was still flying upside-down. Williams attempted to roll back upright; the wing immediately began to fold. Upside-down it was, then.

After quickly running through his options—including running his plane directly into trees to hopefully slow the crash—he opted to remain inverted for as long as possible, then roll back upright at the last possible second, skim the ground on the belly of the plane, "and hope for the best."

To ease his way down, Williams flew a wide, gentle inverted circuit. As he hung there in his harness, with the blood still rushing to his head, he anticipated his next movements. Should he remove the canopy? No, he didn't want the slipstream distorting his height judgment. What if the canopy jammed? In that case, he would have to break his way out; luckily, he knew that the hood of a Zlin Z526 was light enough that it shouldn't be a problem.

As he approached the ground, Williams waited in his seat for the last possible second. Then he acted fast, spinning the plane back upright, feeling the left wing starting to fold up again and only hoping it would hold. Later investigations of the site would show that he'd executed the roll so low, the left wing tip had brushed the ground, and so carefully, the wingtip light was still intact. He felt the plane hit the ground, fearing the whole craft would break apart:

> "I released the controls and concentrated on trying to roll into a ball, knees and feet pulled up and in, and head down protected by arms. I had a blurred impression of the world going past the windscreen sideways and then with a final jolt, everything stopped."

But it wasn't over yet. Clipping free from the harness, he discovered that the gas tanks had split and the canopy had indeed jammed. He struck the canopy free with one hard blow, emerging with only slight bruising and one heck of a story. "It was a nasty experience," he wrote in his official report, with charmingly British understatement, "but a lot can be learned from it...I hope it will never be needed."

What are the odds that anyone else in the same situation would live to tell the tale? It's tough to say. But as the dangerous moments unfolded, Williams's brain went into a higher gear. As the nightmare played out and he struggled for answers, his extensive domain knowledge, mindset and long hours of practice allowed him to do something extraordinary. He was able to leverage his neural processes and tap into flow.

Keep in mind when talking about flow in the face of danger, we're not necessarily talking about flying broken airplanes, but rather encountering the brain's perception of fear. Because the brain is not great with proportionality, fear can overtake us just meeting someone for the first time, or speaking in front of a group at a meeting.

If we approach that meeting with both fear and a growth mindset, we might still be able to experience flow as the net result. By acknowledging the fearful situation and forcing ourselves to push through it, even though we experience discomfort, we can achieve flow in a breakthrough moment of transcendence.

However, if we approach that very same experience out of fixed mindset, the chances of entering flow are negligible, as inner monologue turns negative and we dwell on the risk or possibility of failure. Even for a flying ace like Neil Williams, a fixed mindset might have spelled doom. So, once again, flow starts with attitude and intention.

The Freedom of Flow

People experiencing flow may describe a feeling of complete control over their situation. However, according to Csikszentmihalyi, flow is not about holding onto control, but transcending the fear of losing it. When you let go of anxious second-guessing, you free up more energy for focusing on the task at hand.

Csikszentmihalyi also describes a sense of time passing differently in flow. "Freedom from the tyranny of time does add to the exhilaration we feel during a state of complete involvement," he notes.[366] Deep immersion in a task leaves no room for the mental timekeeping that occupies us when we're less than inspired.

It's interesting to note that the notion of exact, minute-by-minute timekeeping has become much more prevalent in the last 150 years. The rise of train travel instilled in us the need to precisely sync our clocks and watches. Watching the clock tick away as we wait in train stations, airports or boring meetings is, by historical standards, a relatively new concept. There is no mental time keeping during flow.

The Perks of an Autotelic Personality

As Csikszentmihalyi explains, "autotelic" comes from a combination of the Greek words *autos*, "self" and *telos*, "goal." Together, it can be translated as 'to be done for its own sake.' The autotelic person is one who enjoys learning for the sake of learning.

At the annual meeting of the Society for Neuroscience, thousands converge to share their latest research from a wide variety of brain-related experiments. When asked about the practical application of their work, many of these scientists look at me with surprise. They aren't necessarily hoping to revolutionize society or accrue fame or fortune. Rather, they are simply fascinated by how the brain eliminates waste proteins, or what a frog's brain is doing during hibernation.

This is what it means to be autotelic. These scientists are internally motivated. Challenging themselves to solve problems and gain knowledge is its own reward.

In Chapter Two, we met Ron Avitzur, who was driven to finish his 3D math program at Apple, despite the fact that he was no longer getting paid. His flow experience occurred when his passion and work merged. This, and his belief in the product, kept him moving forward.

Autotelic individuals operate largely outside the boundaries of external rewards. This means they tend to be less influenced by material gain. They are more autonomous and self-sufficient. Csikzentmihalyi says that comfort, power, and status are not their end goals.[367] They are perpetually trying to stay in growth mindset, in a constant race of self-improvement. Given the world's demands, it is impossible to sustain growth mindset all the time. However, it is still their default position. And all of these self-driven traits combine in the autotelic person to create a greater than average number of flow experiences.

British philosopher and writer Alan Wilson Watts said, "This is the real secret of life— to be completely engaged with what you are doing in the here and now. And instead of calling it work, realize it is play."[368] For the autotelic person, there is often little distinction between what they do for enjoyment and what they do for employment.

In *Finding Flow*, Csikszentmihalyi notes that many of us don't take the time to put ourselves in learning circumstances that force us to engage new ideas, places, or people.

He writes in part:

> "Our lives are not ours in any meaningful sense; most of what we experience will have been programmed for us. We learn what is supposed to be worth eyeing, what is not; what to remember and what to forget—Through the years, our experience will follow the script written by biology and culture. The only way to take over the ownership of life is by learning to direct psychic energy in line with our own intentions."[369]

We've seen how the rage to master motivates individuals who have dedicated themselves to achieving a goal. The violinists in Anders Ericsson's expertise study from Chapter Four didn't derive joy from all of their practicing, but they did derive a level of satisfaction from making it through the drudgery of training and practice every day.

This is why you can attain a state of flow while doing something as seemingly mundane as painting a fence. Csikszentmihalyi says if you approach a task by breaking it down into the smallest increments and then you pursue each increment "with the care it would take to make a work of art,"[370] your focus and concentration will heighten the experience. It's a matter of intention.

It can be pretty easy to disengage when a task becomes more difficult. Our brains are designed to embrace stasis and conserve energy—great for survival, but not necessarily conducive to higher levels of meaningful experience.

How many of us encounter discomfort and literally make a 180-degree shift in our mental framing, purposely devoting more psychic energy to go deeper, rather than quit the experience? It's not a habit a lot of us have cultivated. Yet Csikszentmihalyi says that digging in is precisely what triggers flow.

Olympic swimmer Michael Phelps trained in the pool for three to six hours every day, for five straight years.[371] Phelps doesn't have super powers, but his years of deliberate practice, domain knowledge, growth mindset and dedication to his goal mean that every time he pulls on his goggles and Speedo, he is opening himself to the possibility of flow. His goal is to perfect each and every stroke, and in a very real way, the reward is in the doing.

As with Phelps, the meaning of an experience lies in how you choose to frame it. If you want to set yourself up for more flow in your life, you'll need to firm up your intention, your determination.

In sports, perhaps more than anything else, we celebrate the refusal to quit. The prospect of witnessing flow draws millions of fans to sporting events every year, and, although people can live vicariously through their favorite sports hero or team, the only place flow is actually happening is on the court, track, field, or pool. To achieve flow, you have to literally be engaged in action, not sitting on the sidelines.

Csikszentmihalyi's research shows that when we engage, we intentionally alter our perception, intensify focus, and fuel our passion. This engagement is the recipe for the most rewarding experiences in our lives. He sees it as fundamental, not just to flow, but to the very notion of happiness.[372]

The Research

How did Csikszentmihalyi come to understand flow—and happiness—in this way? Like any good scientist, he built his theory on the back of experimentation. In order to understand what steps lead to happiness in the human experience, he developed an evaluative system known as Experience Sampling Method, or ESM.[373]

ESM used a programmable watch to send random signals to experiment participants from morning till night. Every time the signal went off, the subject took stock of whom they were with, what they were doing, and what was on their mind. They then assigned a series of number ratings to their "state of consciousness", gauging mood, concentration level, motivation level, self-esteem and so on. Over the years, his team has collected thousands of data sets, creating a sort of pointillist picture of how humans process emotion.

According to Csikszentmihalyi's study, many of us in the US ping-pong between moments of anxiousness and "passive boredom,"[374] even in our leisure time. Television is perhaps the greatest instrument of this passive boredom, since the average American spends a little over five hours a day watching TV.[375] Not surprising, ESM results suggest that happiness is not an outgrowth of the TV experience.

This makes sense: if your central goal is to live a more fulfilled life, you probably won't get there spending roughly five of your 16 daily waking hours in emotional limbo, in front of a glowing screen. Csikszentmihalyi suggests that happiness doesn't come from emotional passivity, but as the byproduct of complete engagement, which brings us back to flow.[376]

Where and How Do We Find Flow?

Flow occurs most often during our free time, when we are engaged in the things we love. But if the conditions are right, it can happen anywhere, including the workplace. The trick is to seek out or construct conditions that allow for peak experience.

For many of us, this requires a willingness to review long-held habits, and to take a fresh look at goals and our mindset as we pursue them. This new perspective can lead to renewed focus, along with a more creative practice strategy. Armed with a keener understanding of how to sustain willpower, and the power of incremental steps, flow becomes the ultimate reward. Csikszentmihalyi says it's worth it: "Flow experience provides the flashes of intense living against the dull background of the daily grind."[377]

Flow Under Pressure

As Steven Kotler hints in the title of his book *Becoming Superman*,[378] true mastery of flow could be the next best thing we have to super powers. Humans may not yet be "more powerful than a locomotive" or "able to leap tall buildings in a single bound," but he explains that when extreme athletes make the most of a flow state, they challenge our very notions of what is possible.

For instance, imagine swimming 446 feet straight down into the ocean—just nine feet short of the Pyramid of Giza's apex—and back up again, on a single breath of air.

As far as mastery over the human brain and body goes, it's hard to get more intense than free diving. People have performed deep dives without equipment for centuries; Japanese tradition suggests that the female *ama* divers were in operation 2000 years ago, swimming to depths of 100 feet to hunt for pearls, seaweed, and shellfish.[379]

However, as an organized sport, free diving has only taken off since the seventies. How do athletes train to hold their breath for over four minutes? How can divers triumph over their own internal organs as their diaphragms begin to contract in panic? Perhaps not surprisingly, for Mandy-Rae Cruickshank, who set the 446-foot world record mentioned above, it all comes down to flow. Even before talking with Kotler and learning about Csikszentmihalyi's theories, Cruickshank was intimately acquainted with the concept.

She told us:

> "I had no idea it had a name. I just thought it was this state you really get into that pisses off all the interviewers. They're trying to find out what you're thinking on a dive. They're always expecting you to say that you're scared when you're under water that deep, or you're looking for sharks, or [noticing] the color of the water around you."

In other words, reporters talk to divers expecting to hear a full summary from analytical System 2. "If you are thinking of that, you're probably not as deep as you should be," Cruickshank explained. In a successful dive, those systems aren't even online.[380]

If that mental state can be hard to verbalize, it certainly isn't difficult for Cruickshank to spot. "You can see it in competitions," she says, "people who have found it, and people that haven't." When asked for the telltale sign of a free-diver in flow, her response is immediate: "They make their dives."

Free-diving demands a tremendous amount from an athlete. In order to stay calm, focused, and kicking on a lungful of air, at depths so extreme that the pressure can spontaneously rupture the human eardrum and the water takes on a thicker, viscous quality from its own collective weight, divers must literally change their physiology.

Cruickshank explains that the average person attempting to hold a breath on a whim will last between 30 and 40 seconds. "And that's not because of a rise in CO_2 or a drop of O_2; it's just that the breathing rhythm has changed." The initial urge to breathe is simply your system upset at the deviation from habit. "If you hold past that, eventually you'll get to the point where the CO_2 is starting to rise." An increase of CO_2 triggers sensors that heat muscles in the diaphragm, to try to force breathing. "And you can still keep working your way past that."

Willpower and a certain comfort with discomfort are definitely part of the process, but just as practicing a skill the right way rewires the brain, practicing free-diving alters the body. Cruickshank recounts taking part in a Simon-Frasier University research project. She and Kirk Krack, her coach and husband, spent twelve weeks teaching their techniques to people with no prior free-diving experience. By the end, the subjects demonstrated an improved blood flow to the brain during breath holds. They had literally changed their physiologies to perform better underwater.

But holding your breath in a pool is one thing; braving the sea can be quite another. "When you're diving in Vancouver waters, below sixty meters, 200 feet, it's black," Cruickshank notes. "It's like you're diving with your eyes closed, even though they're wide open." At this depth, swimming in the ocean can begin to feel like being alone on another planet—an eminently hostile one.

Under such conditions, flow isn't just a perk; it's practically a requirement. Without a seamless unity of mind and muscle, without freedom from the hesitations and rumination of analytical System 2, without focus and finely honed reflexes, without refined proprioception and a sense of inner peace, free diving is nearly impossible. And this means that a free-diver can't just cross their fingers and hope flow happens; the champions drill themselves on getting into this state of mind in the same way they drill themselves on breath holds.

"It's not just an automatic response, at least not in my relation to it. It's part of a lot of mental training," says Cruickshank. The goal is obvious, the challenge is certainly present, but that acute concentration and unflappability can be elusive.

For her, repetition is key to accessing that instinctive calm and control. "When I was doing my records, I had my routine from when I woke up in the morning," she says. Beyond the physical exercises, Cruickshank followed detailed mental routines. Before she even started stretching, she would lie on the ground and mentally run through every aspect of the dive: "A real-time visualization, where you can taste the water and feel everything you're going through on it. Very vivid."

Interestingly, Cruickshank explains that she wasn't lying there envisioning a flawless dive.

> "If you visualize it as being 100% correct, when something does go a little bit off on your real dive, it throws you. So you get to know, sort of, what issues you might have, whether it's your right ear not equalizing as quickly as the left, or different things like that. And you put that into your visualization so that when it does happen on your dive, it's already been thought through and you know how to deal with it, and it doesn't raise your heart rate or anything else. It's part of the plan."

Compare this with Coach Bowman's broken goggles exercise for Olympic swimmer Michael Phelps, or champion pilot Neil Williams flying his broken plane and anticipating the possible need to punch the cockpit canopy open. The goal is to keep everything reflexive and automatic, leaving no room for second-guessing or surprises. In a sport that is all about committing yourself to a bold course, a moment of doubt can be a free-diver's undoing.

Cruikshank has a term for that distracting, self-defeating mental chatter.

> "We call it the Evil Monkey, when you're not focused like that. And if the Evil Monkey has an opening, it will start nattering at you: 'Ooh, that doesn't feel good, I think you need to turn around, you're definitely out of air, something's wrong here.' And you end up with this other voice in your head, you end up turning around and coming up with excuses why you turned early, because there really isn't a reason."

The Evil Monkey has been the downfall of many a free-diver, Cruikshank explains. For Cruikshank, a self-described "basket case," finding a way into that mental calm is a big part of the appeal.

> "Once I get into the water, that's sort of when I let myself become the athlete, and by the time I'm doing my warm-up dives, during that 35 minutes, I just get more and more into what I'm doing, and focusing on just the techniques, so that by the time I'm doing my dives, no one else exists."

The flash of the cameras and the din of the crowd fade away.

> "I've missed my countdown because I've been, just in another world and haven't heard anything that was going on around me."

> "Everything shuts down except for what I need, for what I'm doing. And that allows me to use all the energy just to focus on that. For me, how scattered I normally am with things, to be able to block everything else out is pretty amazing. Like I

was saying at the beginning, it was always the thing that would bug interviewers after a record attempt. They'd be like, 'So, what were you thinking when you were down at 288 feet?' Well, you're not thinking. 'You must've had something going through your mind.'"

If Cruikshank's internal quiet disappoints journalists, it also provides a vivid firsthand account of flow in action:

> "I have this voice that's like a recorder that's going off, counting my kick cycles, reminding me to tuck my chin, relax my stomach, stay streamlined. It's kind of like this ongoing dialogue that's going through a checklist. And absolutely nothing else goes through my mind. Every once in a while, when I grab tag at bottom, I have a split second where I smile to myself and think, 'I hit world record depth!' And then you go back into that zone, and everything else disappears again."

Perhaps not surprising for an athlete who has broken multiple world records, Cruikshank considers herself a competitive person. At the same time, she displays a true autotelic love of the sport. Free diving is "good for the soul," she told us:

> "For me, it's a very calming thing, and it centers me more, and shows me that there's more purpose. Like, it's nice running a company and raising a daughter and everything, it all has its goals, but it's nice to do something that's just for you. If I'm not doing a world record or a national record, it's not benefiting sponsors or anything like that; it's just benefiting me, that I know I can still pull this off. And experience that nothing-exists zone. And just relax and have fun with it."

Maybe tugging on a wetsuit and imitating a submarine is not your idea of fun, but Cruikshank has found her own personal pathway to flow, and for her, the answer lies in the bottom of the ocean. "It's just so peaceful down there," she says. "That's all I can say about it...It's very, very calm and peaceful."

Of course, as Cruikshank proves, there is no free lunch to accessing that kind of peace. Getting into increased states of flow requires work. One of the keys to that work is improving your focus. Stanford's Dr. Kelly McGonigal from Chapter Three has been instrumental in showing how meditation can improve both willpower and focus.

Focus: A Gateway to Flow

There are three forms of meditation in common practice today: focused attention, mindfulness and compassionate meditation.[381] All three forms are extremely beneficial, but we'll explore focused attention because it is more closely aligned to gearing up willpower and flow state.

People have performed meditative activities for thousands of years, across many religions, but in this context, we're not talking about a spiritual practice. We're talking about meditation in terms of cognitive science. According to the National Center of Complementary and Alternative Medicines, part of the National Institutes of Health, meditation is one of the top ten alternative health therapies used in the U.S.[382]

Many studies can speak to the associated benefits. A study in the journal of Psychological Science reported that after only 10 minutes of meditative practice a day during a two week clinical trial, a randomized group of undergraduate students showed a 16-percentile point increase on the GRE, as well as improved working memory.[383]

Meditation taps into your executive control center and its subsystems, allowing you to quiet the urges to yield to instant gratification—to eat that marshmallow—as well as keeping you more focused on a given task. Regular meditation will improve the oxygen and blood flow to your brain,[384] help you to make better-quality decisions, and enhance your willpower throughout the day. And all that predisposes you toward a flow experience.

The Essence of Meditation

When researcher Wendy Hasenkamp was at Emory University, she and her team ran experiments to understand what happened to the brain during focused meditation.[385]

Once the test subjects entered the fMRI scanner, they were told to focus on the sensation of their own breathing. They were instructed to push a button when they noticed that they were no longer concentrating on their breathing.

In the end, the researchers came to recognize a distinct pattern among their subjects, involving four phases and four distinct brain areas. The subjects' default-mode was mind wandering. As we've discussed in Chapter Two, this helps conserve mental energy.

In phase two, when the subjects tried to refocus on their breathing, their salience network would take over. This is the part of the brain that registers sudden attention shifts as you become aware of distractions happening around you. Your salience network might be more aptly named your distraction network, and for many of us, this network is frequently on high alert.

As we learned in Chapter Two, our concentration is highly susceptible to surprising sights and sounds. Try staying focused as a police car screeches past you, siren blaring and lights flashing. Even without the police car, just attempting to stay focused on your own breathing—something you don't regularly do—becomes a challenge, especially when you're crammed inside an fMRI.

In the third phase, the test subject would attempt to stop paying attention to the distraction and wrestle focus back to breathing. Finally, in the last stage, the prefrontal cortex would reestablish its dominance and refocus on breathing.

Regardless of whether we're talking about an experiment on breath concentration, or just living our daily lives, this four-stage process of focusing, mind wandering, letting go of the distraction, and reestablishing focus is a ritual that we practice all day long. The ease with which you can recapture focus has enormous implications for everything you do, particularly any goal, especially when trying to maintain willpower or enter flow.

Hasenkamp cites a study by Matthew Killingsworth and Daniel Gilbert, in which the team surveyed over 2,000 adults across the day, finding that roughly 47% of the time, they weren't concentrating on their current actions. "Even more striking," she adds, "when people's minds were wandering, they reported being less happy." This coincides with Mihály Csikszentmihalyi's research that happiness requires focus and commitment.[387]

Interestingly, University of Wisconsin researchers found that people with at least 10,000 hours of meditative practice under their belt showed less activation in their attention-related brain regions than novice meditators.[388]

The study suggests that once a person has mastered meditation, they need less energy to maintain an attentive state—they become more focus efficient. In other words, meditation, like any other skill, follows Hebb's rule—the more you use a particular neural circuit, the stronger it gets. The bottom line is that meditative practice improves willpower, and improved willpower is a key to maintaining and growing focus.[389] Focus is the gateway to flow.

Meditation How-To's

Writing for Headspace.com, meditation expert Andy Puddicombe outlines a simple meditation exercise. The first step is to find a comfortable position. You can sit upright in a chair or cross-legged on the floor—whatever feels natural. You can even lie down, although you may find yourself fighting the urge to fall asleep. You can be walking in the woods, on a garden path, or even navigating a hallway in your house—any area that offers a relaxed setting with minimum distraction will work.

The next step is to be mindful of your breathing. Simply relax your breathing and direct your focus toward a particular point.

The point could be on a physical object, like a vase in front of you, or a scene in your mind, like the ocean tide rushing up the beach. It can be a repetitive chant, or it can be a slow, repeating count to ten. It can be your next inhalation or footfall. The important thing is to relax and identify a point to direct your attention. Here are a few hints to keep in mind:

1. Remember, your goal is to relax your body and practice maintaining your focused attention on a single target.

2. Find a quiet time in the day when you'll be uninterrupted, and try to stick to the same time. Start with a few minutes a day, with an eventual goal of 20 minutes, if possible.

4. Maintaining true focus can be much harder than it sounds. Especially in the beginning, you may find yourself struggling with frequent distracting thoughts. The key is not to judge your progress, but to pursue your goal with a growth mindset. Give yourself credit for each attempt.

5. Meditation is not a race, and there are no winners or losers. The more you do it, the more you'll set up your brain for better focus. Keep in mind you have the rest of your life to perfect your meditative skill.[390]

It's easy to get frustrated. Someone who didn't stick with meditation will tell you, "I tried to imagine myself sitting in a serene beach scene, and it didn't work. My mind just kept wandering to 'Oh, I've got to pick up Susan from dance' or 'What am I making for dinner?' or 'Wow, I am really not concentrating very well.' I just couldn't keep it going."

However, McGonigal says, here's what they don't realize: those moments where your attention meanders away and you force it back, those instances are what actually strengthen your meditating abilities. It's like lifting weights: the more reps you do, the more powerful the muscle becomes.[391]

It's the opposite of how you might think it would work. As you're sitting there struggling to focus, snapping yourself back from your default mind wandering, that struggle marks your real progress. You are increasing the efficiency of the four steps identified by researcher Wendy Hasenkamp and her team: focusing, mind wandering, letting go of the distraction, and reestablishing focus. And more focus means more willpower. It sounds too easy, but that's the beauty: it really is that easy. The important thing is to make time in the day to meditate.

In fact, meditation has an interesting side effect: many people who meditate experience up to an additional hour of sleep at night.[392] As discussed in Chapter Three, this additional hour of rest has all kinds of health benefits, including lowering your general level of anxiety.

If you're not into the idea of meditation, you can take solace in the fact that regular exercise can have a similar effect on the brain. However, if you want to build maximum focus and make the most of your mental potential, nothing beats the combination of exercise and meditation.[393] And once your focus power increases, it often spills over into other areas of everyday living, making it easier to avoid *procrastination roundabout* and the siren call of Future-Self. Best of all, it becomes easier to achieve flow.

Chapter Key Points: Hacking Your Flow App

- Having established deep domain knowledge through deliberate practice, flow involves intense focus while pushing yourself to the edge of your abilities.

- Flow is a reflexive System 1 enterprise. Action and awareness merge to create a state of hypofrontality, operating beyond System 2's executive control.

- To achieve flow, break a physical activity into small and manageable increments, and then pursue each increment "with the care it would take to make a work of art."

- Flow is the exact opposite of "simply going through the motions."

- Think of meditation as a gateway to flow

- There is a direct correlation between flow and personal fulfillment.

Putting It All Together:

The Maestro of Brain Apps

To illuminate the true potential of all seven brain apps working in concert, let's travel back in time again. Our destination is no longer La Guardia High School in 1995, but fifteenth century Italy. Here, in the town of Vinci, we encounter the illegitimate son of a notary and a local peasant girl. It is not an auspicious beginning, and yet young Leonardo Da Vinci would go on to create two of the world's most iconic paintings, the *Mona Lisa* and *The Last Supper*.

That alone might earn him a place in the pantheon of all-time great artists, but he was also a world-class sculptor, musician, engineer, anatomist, and inventor. His sketchpads were filled with designs for helicopters, parachutes, paddleboats, and even a repeating rifle—hundreds of years before any of these would come into existence.

By the time he died at age 67, he was considered to be one of the greatest minds of his century. For many, Leonardo Da Vinci is the essential 'Renaissance Man,' his name synonymous with genius. In *How to Think like Da Vinci*, author and Da Vinci scholar Michael J. Gelb writes:

> "Leonardo developed astonishing powers of sight bordering on those of a cartoon superhero. In his 'Codex on the Flight of Birds,' for example, he recorded minutiae about the movements of feathers and wings in flight that remained unconfirmed and not fully appreciated until the development of slow-motion moving pictures."[394]

When we talk about someone with so many achievements, it is easy to mythologize, to place that individual above the rest of us mere mortals. But many scholars, from Linda Nochlin to Anders Ericsson, are quick to remind us that genius is always grounded in practice.

Indeed, Gelb says the Maestro's secret to success is no secret at all. He intentionally cultivated specific habits that catapulted his abilities to the very pinnacle of virtuosity.[395] Gelb's observations about Da Vinci's life correlate strongly with the description of our brain apps.

In 1472, at the age of twenty, Da Vinci was invited into the prestigious Guild of Florence, an exclusive group of artists. He turned them down, instead opting to continue studying with his tutor, the artist Andrea del Verrocchio of Florence, until Da Vinci himself was satisfied that he had sufficiently honed his skills and created powerful sustainable habits.[396] In other words, he was less interested in personal glory than he was in the autotelic work of refining his craft—and in following his rage to master to a level of peak performance.

Mindset

In Chapter One we discussed how growth mindset involves an emphasis on learning, taking risks, and pushing to the edge of your comfort zone. A growth mindset embraces challenges and persists in the face of obstacles and failure. This was certainly true of Da Vinci, who fully immersed himself in the specifics of whatever domain captured his attention. His willingness to experiment with new ideas and risk failure was fundamental to his modus operandi—classic growth mindset at work.

Da Vinci's hunger for knowledge and improvement led him to sometimes take enormous risks. For instance, he was said to have engaged in grave robbery; the study of dissecting cadavers was one of the few ways to become an expert in human anatomy. Perhaps not surprisingly, digging up corpses was against the law, and could result in prison time. But when it came to understanding the intricacies of the human form, as with so many other subjects, Da Vinci refused to settle for mere speculation.

His creative problem solving was not stymied by societal conventions; grave robbing was a crime. Still, the results speak for themselves. His anatomical drawings of the human spine are of such exceptional caliber, they're still used in medical books today.[397]

Some of Da Vinci's experiments ended in disaster. In 1495 and 1496, while working on a painting for the refectory of Santa Maria dell Grazie, Da Vinci tried out what is believed to be a form of egg-based tempera and oil paint, a new technique at the time.[398] Due to his use of this unproven method, one of his greatest works, *The Last Supper*, began to fall apart before it was finished.

Since its inception, the painting has required numerous rounds of restoration. Today, the debate continues over how much of the original painting remains intact and how much has been retouched. Even for Da Vinci, growth sometimes meant learning from embarrassing mistakes.

Much to the chagrin of his many patrons, Da Vinci did not feel bound by contract or time. His drive for perfection and his desire to produce the finest work possible superseded his adherence to specific deadlines. Again, in true autotelic style, money was not his primary motivator. Despite his fame, at no point in his life did he ever amass personal fortune. This didn't seem to matter much to him; his true goal seems to have been simply to create and to learn.

He was so drawn to the idea of increasing his domain knowledge that he was known to sign his writing as "Disciple of Experience."[399]

Goal Strategy

In Chapter Two, we suggested that making significant progress toward any goal requires a premeditated strategy. DaVinci was not acquainted with the **SMARTR** system, which entails examining your goal intentions through the checklist of specific, measurable, actionable, realistic, time-oriented and reward-based outcomes. Instead, Da Vinci adhered to what would later become known as the scientific method.

With the scientific method a problem is identified, data is gathered, a hypothesis is formed, the hypothesis is tested, the results are examined, and either the hypothesis is confirmed or amended. Purposefully employing a problem solving strategy for everything from water management to siege technology allowed DaVinci's ideas to outlast many his famous peers of the day.

Although outcomes are never certain, following a systematic strategy for exploration as Da Vinci did allows for consistency and improves the chances of repeatable success.

DaVinci recorded his hypotheses, tests, results and observations in a series of leather bound notebooks today referred to as the 'codices.' These manuscripts contained thousands of pages of notes and illustrations organized around four general themes—mechanics, painting, architecture, and human anatomy.[400] Although some of the codices have been lost to time, this is one of the principal ways we know what we do today about Da Vinci's inventive mind and the many experiments he conducted.

(In 1994, Bill Gates paid an estimated $30.8 million for one of Da Vinci's notebooks, the *Codex of Leicester*, making it the single most valuable manuscript in the world.[401])

Willpower

As we outlined in Chapter Three, the ability to display fortitude and delay gratification is the best indicator of future success. Increasing fortitude requires willpower's essential three building blocks: exercise, sleep and nutrition. Writings in his 500-year-old *Codex Alanticus* suggest Da Vinci's lifestyle was ahead of its time. He carefully tended to his willpower triad.

During his experiments in human dissection, Da Vinci had discovered clogged arteries in some of the cadavers, which he attributed to their diet. Perhaps this was the impetus of his lifelong vegetarianism, in an era when meat was a symbol of prosperity and success. Writer Alex Naidus notes that Da Vinci's other health tips include proper rest, avoiding overeating, and drinking only moderate amounts of wine.

Centuries before the health craze of the 1990's, Da Vinci was also a proponent of physical fitness. He was said to have been an excellent horseman and enjoyed a wide range of physical activates, including hiking through the Italian countryside of his beloved Tuscany.[402] And on those hikes, Da Vinci conducted many thought experiments, devising workable designs for contraptions like the bicycle, submarine and military tank.

None of these would come to fruition in his lifetime, largely because his creativity outpaced the materials technology of his day. Still, by maintaining a healthy lifestyle in an era where few understood the ramifications of a lack of exercise, poor diet and poor sleep, Da Vinci was able to live longer than most. This allowed him to extend his creative focus and output well into his later years.

Perhaps nothing speaks more to Da Vinci's ability to delay gratification than his continued quest for perfection. In his early years from his refusal to leave the tutelage of his master until he was convinced he had mastered key techniques, to spending years working on projects, not willing to concede the work was done until he had satisfied himself there was nothing more he could do.

Deliberate Practice

In Chapter Six, we talked about how the Renaissance was partly enabled by wealthy patrons like the Medicis, whose funding and support connected some of the best minds of the era. However, a number of other factors helped bring those artistic achievements into being, including, as Linda Nochlin has noted, an apprenticeship program that treated fine arts like any other trade to be learned through years of intense study.

It seems Da Vinci knew instinctively what researcher Anders Ericsson proved centuries later: domain expertise comes down to accumulating the right kind of deliberate practice under the right conditions. Of course, no one was talking about neural pathways or myelin insulation 500 years ago, but dedicated craftspeople understood the connection between practice and expertise.

At the age of fifteen, Da Vinci's father had apprenticed him to well-known painter and sculptor Andrea del Verrocchio. For the next decade, Da Vinci was tutored in a wide variety of art forms, from painting to sculpting to goldsmithing to the mechanical arts. During this seminal ten-year stint, Da Vinci was not practicing and perfecting his skills in a vacuum. He was always competing against the best and brightest of equally determined apprentices—perhaps the ultimate feedback loop for continuous incremental improvement.

As Frans Johansson explained, this hotbed of competition and idea exchange was the perfect environment to foster radical growth. Da Vinci was in the right place at the right time, under the right circumstances for an explosion of intersectional ideas to take hold.

Habit Formation

What makes Da Vinci's painting so remarkable, in part, was his dedication to detail.[403] Throughout his life he cultivated specific habits in everything from his lifestyle, to his work ethic, to his problem solving strategy to the specificity of how he pursued the arts. Adhering to the scientific method taught him the value of repeatable outcomes, which in turn led him to understanding the power of habit in the pursuit of exceptional quality.

This was particularly true in his painting techniques. He painstakingly developed a series of custom glazes that allowed for intricate shading in a process called *chiaroscuro*. He worked incessantly until he'd habituated this technique as an integral aspect of his painting style. Chiaroscuro permitted him to transition from one color to the next, blending them so seamlessly that it's impossible to tell where one color begins and the other ends.

Among his most celebrated painting techniques, and one he used to great effect on the Mona Lisa, is *sfumato*, which creates a smoky haziness, giving his subject an otherworldly quality. It's part of what makes the Mona Lisa's smile so enigmatic. He achieved this through habitually layering ultra-fine layer upon ultra-fine layer of nearly translucent finish over a period of years, a tedious but brilliant strategy to work around the technological limitations of his time.

Creativity

As we discussed in Chapter Six, a central theme in creative endeavors is analogous thinking, using ideas from one domain to seed a solution in another domain. Da Vinci frequently engaged his analytical System 2 process, priding himself on conducting experiments, testing scientific theories, and finding novel solutions to the problems of his day by applying concepts he'd learned through his multifaceted studies. He was equal parts artist and scientist.

If the combination of artist-scientist seems strange today, it made perfect sense to Da Vinci. Gelb says Da Vinci understood art and science not as two separate systems, but as two halves of the same coin.[404] His anatomy studies, his interest in the natural world, and his chemistry experiments with paints and finishes gave him a uniquely varied range of inspirations and recombinations to make Kirby Ferguson proud. Copy, transform and combine were at the heart of much of Da Vinci's work.

His knowledge of engineering, in conjunction with his accomplished musicianship, enabled him to design and create new instruments—a lyre cast in solid silver in the shape of a horsehead, reported to be superior to any of the wooden lyres of his day,[405] as well as plans for a 'viola organista', a sort of mash-up between a harpsichord and a viola.[406] Deep domain knowledge in a variety of fields gave Da Vinci the ability to transfer ideas and technologies across domains. This transference, or lateral thinking, allowed Da Vinci to repurpose mechanics for war machinery into a variety of applications, including water movement and lifting devices.

Flow

Central to the idea of Flow in Chapter Seven is pursuing a task with such intense concentration and passion that your actions and perception merge. Time is no longer relevant and one's total being is engulfed in the activity at hand.

Consider how art scholars at the Louvre believe that Da Vinci labored for nearly twelve years on his 30-inch tall tour de force, the Mona Lisa.[407] Da Vinci was driven by the need to meticulously ensure that every single brush stroke served the final desired outcome: a transformative effect on the viewer. For Da Vinci, the progression of time itself took a backseat to the act of perfecting his painting. His refined skill, coupled with his ability to focus at such a micro level, suggests he was frequently operating in a state of flow.

The result: extraordinary achievements that still captivate us five centuries later. His success was not a series of lucky accidents, but instead, the fruit of intense domain knowledge and a well-orchestrated life.

Conclusion

Mark Twain's best friend, writer and critic Dean Howells, once spoke of Twain's unique access to "that mystical chaos, that divine ragbag which we call the mind."[408] That was over a hundred years ago. Today we can study and harness the workings of that 'divine ragbag' in ways that would probably astound Samuel L. Clemens.

Across a wide spectrum of domains—Dave Guy's trumpet playing, Ron Avitzur's coding, the Polgar sisters' chess triumphs, Scottie Olsen's inventions, Lin-Manuel Miranda's rap opera pioneering, Mandy Rae Cruikshank's extreme feats of free diving, and finally, the timeless art of Leonardo Da Vinci—we've seen the exceptional effects of leveraging the brain's neuroplasticity in pursuit of a goal.

There is no one secret for joy or success, but you can unquestionably set yourself up for more positive outcomes by developing brain apps for growth mindset, prioritizing your goals, strengthening your willpower, sticking to a deliberate practice regimen, being intentional about your habits, nurturing your creativity, and acting in the pursuit of flow.

The way forward has been made clear; the next adventure is up to you. Good luck, and remember to watch out for the Evil Monkey.

May the flow be with you!

Acknowledgments

It turns out that writing a book is not a solitary endeavor. I begin with a long and persistent shout out to Mary, who through the years has been kind enough to endure my nonstop prattling of ideas. Your incredible patience, feedback, and keen sensibilities have, over and over again, proven invaluable to this manuscript, and my life.

A debt of gratitude to Jessica for her dogged determinedness, energy, wit, research prowess, and above all, incredible writing and editing.

Jimmy "The junkman" D, Kevin "Cabinets" M, Sarah L, Marcy B, and of course the redoubtable Consuming Monk—I appreciate your willingness to scale mountains of text to create a better reading experience.

Lucy "*Ka*-Pow," thanks for your design creativity, all purpose council and "one minute to air" bailouts.

Sammy 3, your detailed analysis and nonstop musicality have enriched the read and my brief time on this planet.

Freemie, thank you for your timely assessment, your input on the title, and more importantly, your lifelong friendship.

Casey, lending your voice to this project made all the difference. Your audio read creates an entirely new experience.

Taylormade, thanks for your valuable ear and chill vibe.

Lauren, Jonesy, Fia, Janisimo, Southern Bests, Gleason and Robinson clans, Team Freeman, Powers Inc, Giamby's, Mango, Clarko, Big T, Reilly gals, Wen, Lillian, Manisha, Special K, Erica T, Joe D, Ethan, Silky B, L.G., Sid "The Kid," Jack S, Tim J and the fine folks at Elkay—you are at the molten core of Best Nation.

ACKNOWLEDGMENTS

Special thanks to my agent, Famous Dave (or Donnie "if you're into the whole brevity thing.")

This book is dedicated to Sweet Phyllis and the Big G. Your support has been an unending Mobius strip of unconditional love.

Endnotes

Chapter One

[1] Swaminathan, Nikhil. "Why Does the Brain Need So Much Power?" *Scientific American*, 29 Apr. 2011, www.scientificamerican.com/article/why-does-the-brain-need-s/.

[2] Kahneman, Daniel. *Thinking, Fast and Slow*. New York: Farrar, Straus and Giroux, 2011.

[3] Williams, Richard. "January 1, 1925: Cecilia Payne-Gaposchkin and the Day the Universe Changed." *APS News*, Jan. 2015, www.aps.org/publications/apsnews/201501/physicshistory.cfm.

[4] Gorman, Sara E., Gorman, Jack M. *Denying to the Grave: Why We Ignore The Facts That Will Save Us*. Oxford University Press. USA. 2016

[5] Isaacson, Walter. *Einstein: His Life and Universe*. New York: Simon & Schuster Paperbacks, 2007.

[6] Farr, Emma-Victoria. "JK Rowling: 10 facts about the writer." *The Telegraph*, 27 Sep. 2010, www.telegraph.co.uk/culture/books/booknews/9564894/JK-Rowling-10-facts-about-the-writer.html.

[7] Gladwell, Malcolm. *Outliers: The Story of Success*. New York: Little, Brown, 2008.

[8] Gladwell, Malcolm. *Outliers: The Story of Success*. New York: Little, Brown, 2008.

[9] Deaner, Robert O., Aaron Lowen, Stephen Cobley. "Born at the Wrong Time: Selection Bias in the NHL Draft." 27 Feb. 2013. *PLoS ONE*, 10.1371/journal.pone.0057753.

[10] Dweck, Carol S. *Mindset: The New Psychology of Success.* New York: Random House, 2006.

[11] Sweeny, Alastair. *BlackBerry Planet: The Story of Research in Motion and the Little Device That Took the World by Storm.* Mississauga, Ont.: John Wiley & Sons Canada, 2009.

[12] Wheaton, Ken. "Dubious Distinction: Crackberry Is Word of the Year." *Advertising Age*, 1 Nov. 2006, adage.com/article/adages/dubious-distinction-crackberry-word-year/112906.

[13] Melanson, Donald. "Fortune names RIM fastest growing company… in the world." Engadget, 18 Aug. 2009, www.engadget.com/2009/08/18/fortune-names-rim-fastest-growing-company-in-the-world.

[14] Hicks, Jesse. "Research, no motion: How the BlackBerry CEOs lost an empire." *The Verge*, 21 Feb. 2012, www.theverge.com/2012/2/21/2789676/rim-blackberry-mike-lazaridis-jim-balsillie-lost-empire.

Gustin, Sam. "The Fatal Mistake That Doomed BlackBerry." *TIME.com*, 24 Sep. 2013, business.time.com/2013/09/24/the-fatal-mistake-that-doomed-blackberry.

F., Alan. "They said what? Great quotes from Jim Balsillie and Mike Lazaridis." *phoneArena.com*, 29 Jun. 2012, www.phonearena.com/news/They-said-what-Great-quotes-from-Jim-Balsillie-and-Mike-Lazaridis_id31751.

Gustin, Sam. "The Fatal Mistake That Doomed BlackBerry." *TIME. com*, 24 Sep. 2013, business.time.com/2013/09/24/the-fatal-mistake-that-doomed-blackberry.

[15] Arthur, Charles. "RIM chiefs Mike Lazaridis and Jim Balsillie's best quotes." *The Guardian*, 29 Jun. 2012, www.theguardian.com/technology/2012/jun/29/rim-chiefs-best-quotes.

[16] Rodier, Melanie. "How RIM Murdered BlackBerry." *Wall-Street & Technology*, 29 Jun. 2012, www.wallstreetandtech.com/asset-management/how-rim-murdered-blackberry/d/d-id/1266489?.

[17] Hicks, Jesse. "Research, no motion: How the BlackBerry CEOs lost an empire." *The Verge*, 21 Feb. 2012, www.theverge.com/2012/2/21/2789676/rim-blackberry-mike-lazaridis-jim-balsillie-lost-empire.

[18] "comScore Reports April 2015 U.S. Smartphone Subscriber Market Share." *comScore.com*, 5 Jun. 2015, www.comscore.com/Insights/Market-Rankings/comScore-Reports-April-2015-U.S.-Smartphone-Subscriber-Market-Share.

[19] Adams, Susan. "The Worst CEO Screw-Ups Of 2013." *Forbes*, 18 Dec. 2013, www.forbes.com/sites/susanadams/2013/12/18/the-worst-ceo-screw-ups-of-2013/#4df78e9661c0.

[20] Dweck, Carol S. *Mindset: The New Psychology of Success*. New York: Random House, 2006.

[21] Hanson, Rick, and Richard Mendius. *Buddha's Brain: The Practical Neuroscience of Happiness, Love & Wisdom*. Oakland, CA: New Harbinger Publications, 2009.

[22] Whitley, Carla Jean. *Muscle Shoals Sound Studios: How the Swampers Changed American Music*. Charleston, SC: The History Press, 2014.

[23] Edison Innovation Foundation, "Edison Patents." 2014, www.thomasedison.org/index.php/education/edison-patents.

[24] Uth, Robert, and Margaret Cheney. "Tesla: Life and Legacy." *PBS.org*, 2000, www.pbs.org/tesla/ll/ll_america.html.

[25] Hendry, Erica R. "7 Epic Fails Brought to You By the Genius Mind of Thomas Edison." *Smithsonian.com*, 20 Nov. 2013, smithsonianmag.com/innovation/7-epic-fails-brought-to-you-by-the-genius-mind-of-thomas-edison-180947786.

[26] DeGraaf, Leonard. *Edison and the Rise of Innovation*. New York: Sterling Pub Co, 2013.

[27] Nochlin, Linda. "Why Have There Been No Great Women Artists?" (1971). *Women, Art, and Power and Other Essays*. Boulder, CO: Westview Press, 1988, pp. 145-178.

[28] "Roger Bannister breaks four-minutes mile." *A + E Networks*, 2009, history.com/this-day-in-history/roger-bannister-breaks-four-minutes-mile.

[29] Burfoot, Amby. "How To Set a World Record in the Mile." *Runner's World*, 13 Feb. 2014, www.runnersworld.com/newswire/how-to-set-a-world-record-in-the-mile.

[30] "About: Bill Bowerman." *U.S. Track and Field Coaches Association*, 2016, www.ustfccca.org/ustfccca-awards/the-bowerman/bill-bowerman.

[31] Moore, Kenny. *Bowerman and the Men of Oregon: The Story of Oregon's Legendary Coach and Nike's Cofounder*. Emmaus, PA: Rodale, 2006.

[32] Corasaniti, Mike. "The End of an Era: Steve Prefontaine's Last Major Record Falls." *Bleacher Report*, 3 Jul. 2012, bleacherreport.com/articles/1245002-the-end-of-an-era-steve-prefontaines-last-major-record-falls.

[33] Bowerman, William J. and W.E. Harris. *Jogging: A Physical Fitness Program for All Ages*. Ace Books, August 1967

[34] Mathew, Alicia. "Nike, Inc. Success Story." *SuccessStory.com*, successstory.com/companies/nike-inc.

Danani, Suman. "NIKE – History & Heritage." *Academia.edu,* www.academia.edu/3679201/NIKE_history_and_heritage_SUMAN.

[35] "Bill Bowerman." *University of Oregon,* sportshistory.uoregon.edu/topics/the-track-and-field-legacy/famous-coaches/bill-bowerman.

[36] "Nike Mission Statement." *Nike,* help-en-us.nike.com/app/answers/detail/a_id/113/~/nike-mission-statement.

[37] Byrne, John. "CEO Disease." *Bloomberg,* 31 Mar. 1991, http://www.bloomberg.com/news/articles/1991-03-31/ceo-disease.

[38] Tsai, Allen. "The True Story Behind the Rise and Fall of BlackBerry." *2machines*, Aug. 2013, 2machines.com/184127.

[39] DiCarlo, Lisa. "How Lou Gerstner Got IBM To Dance." *Forbes,* 11 Nov. 2002, www.forbes.com/2002/11/11/cx_ld_1112gerstner.html.

[40] Vollmer, Lisa. "Anne Mulcahy: The Keys to Turnaround at Xerox." *Stanford Business*, 1 Dec. 2004, www.gsb.stanford.edu/insights/anne-mulcahy-keys-turnaround-xerox.

George, Bill. "America's Best Leaders: Anne Mulcahy, Xerox CEO." *US News & World Report*, 19 Nov. 2008, www.usnews.com/news/best-leaders/articles/2008/11/19/americas-best-leaders-anne-mulcahy-xerox-ceo.

[41] Ainslie, Scott. *Robert Johnson*. S.l.: Hal Leonard, 2005.

Chapter Two

[42] Avitzur, Ron. "The Graphing Calculator Story." 2004, www.pacifict.com/Story.

[43] Drucker, Peter F. *The Practice of Management*. New York, NY: Collins, 2006.

[44] Doran, G. T. "There's a S.M.A.R.T. Way to Write Management's Goals and Objectives", *Management Review*, Vol. 70, Issue 11, Nov. 1981, pp. 35-36.

[45] Young, Scott H. "Two Types of Growth." *ScottHYoung.com*, Feb. 2013, www.scotthyoung.com/blog/2013/02/05/two-types-of-growth.

[46] Drummond, Helga. *Guide to Decision Making: Getting It More Right than Wrong*. Hoboken, NJ: John Wiley & Sons, 2012. pp. 126.

[47] B.R., "A crash course in probability." *The Economist*, 29 Jan. 2015, www.economist.com/blogs/gulliver/2015/01/air-safety.

[48] Gawande, Atul. *The Checklist Manifesto: How to Get Things Right*. New York: Metropolitan, 2010.

[49] Achor, Shawn. "When a Vacation Reduces Stress — And When It Doesn't." *Harvard Business Review*, 14 Feb. 2014, hbr. org/2014/02/when-a-vacation-reduces-stress-and-when-it-doesn't.

[50] Levitin, Daniel J. *The Organized Mind: Thinking Straight in the Age of Information Overload*. London: Viking, 2014. Kindle edition.

[51] Allen, David. *Getting Things Done: The Art of Stress-free Productivity*. New York: Penguin, 2001.

[52] Sherman, Erik. "The Legendary Time-Management Tip That's Worth $550,000." *Inc.com*, 7 Jan. 2013, www.inc. com/erik-sherman/the-legendary-time-management-tip-thats-worth-$550000.html.

[53] Iqbal, Shamsi T. and Eric Horvitz. "Disruption and Recovery of Computing Tasks: Field Study, Analysis, and Directions." *Proceedings of ACM CHI 2007 Conference on Human Factors in Computing Systems*, 2007, pp. 677-686, research.microsoft.com/en-us/um/people/horvitz/chi_2007_iqbal_horvitz.pdf.

[54] Levitin, Daniel J. *The Organized Mind: Thinking Straight in the Age of Information Overload*. London: Viking, 2014. Kindle Edition.

[55] Menon, Vinod, and L.Q. Udin. "Saliency, switching, attention and control: A network model of insula function." *Brain Structure and Function*, Jun. 2010, www.ncbi.nlm.nih.gov/pmc/articles/PMC2899886.

Levitin, Daniel J. *The Organized Mind: Thinking Straight in the Age of Information Overload*. London: Viking, 2014. Kindle Edition.

[56] Lawrence, John, and Dan Hise. *Faulkner's Rowan Oak*. Jackson: U of Mississippi, 1993. pp. 26.

[57] Herbert, Geoff. "iPhone 6S fans line up at Apple Store, prepared to wait all night in Destiny USA." *Syracuse.com*, 24 Sep. 2015, www.syracuse.com/business-news/index.ssf/2015/09/iphone_6s_lines_apple_store_syracuse_destiny_usa.html.

[58] "Accident Overview." *Federal Aviation Administration*, lessonslearned.faa.gov/ll_main.cfm?TabID=3&LLID=8&LLTypeID=2.

[59] Hallinan, Joseph T. *Why We Make Mistakes: How We Look Without Seeing, Forget Things in Seconds, and Are All Pretty Sure We Are Way Above Average*. New York: Broadway, 2009. pp. 78.

[60] Moore, Stephanie. "Rebate Madness - How to Avoid the Rebate Trap." *Consumer Affairs*, www.consumeraffairs.com/consumerism/rebate_madness01.html.

[61] Stelter, Brian. "8 Hours a Day Spent on Screens, Study Finds." *Gainesville Sun*, 27 Mar. 2009, www.gainesville.com/lifestyle/20090327/8-hours-a-day-spent-on-screens-study-finds.

[62] Fallows, James. "Linda Stone on Maintaining Focus in a Maddeningly Distractive World." *The Atlantic*, 23 May 2013, www.theatlantic.com/national/archive/2013/05/linda-stone-on-maintaining-focus-in-a-maddeningly-distractive-world/276201.

[63] Ghose, Tia. "Why the Internet Sucks You in Like a Black Hole." *Live Science*, 23 May 2013, www.livescience.com/34649-why-internet-is-addictive.html.

[64] Csikszentmihalyi, Mihaly. *Flow: The Psychology of Optimal Experience*. New York: Harper & Row, 1990.

[65] "Spaced Learning." *Innovation Unit,* www.innovationunit. org/sites/default/files/Spaced_Learning-downloadable_1.pdf.

[66] Clement, Scott. "America's love/hate (but mostly hate) relationship with taxes, in 7 charts." *The Washington Post*, 15 Apr. 2014, www.washingtonpost.com/news/the-fix/wp/2014/04/15/americas-lovehate-but-mostly-hate-relationship-with-taxes-in-7-charts.

Chapter Three

[67] Twomey, Steve. "Phineas Gage: Neuroscience's Most Famous Patient." *Smithsonian.com*, Jan. 2010, www.smithsonian-mag.com/history/phineas-gage-neurosciences-most-famous-patient-11390067/?no-ist.

[68] Nutt, Amy Ellis. *Shadows Bright as Glass: The Remarkable Story of One Man's Journey from Brain Trauma to Artistic Triumph*. New York: Free, 2011.

[69] Twomey, Steve. "Phineas Gage: Neuroscience's Most Famous Patient." Smithsonian.com, Jan. 2010, www.smithsonian-mag.com/history/phineas-gage-neurosciences-most-famous-patient-11390067/?no-ist.

[70] Kahneman, Daniel. *Thinking, Fast and Slow*. New York: Farrar, Straus and Giroux, 2011.

[71] Berardino, Mike. "Mike Tyson explains one of his most famous quotes." *Sun Sentinel*, 9 Nov. 2012, articles.sun-sentinel. com/2012-11-09/sports/sfl-mike-tyson-explains-one-of-his-most-famous-quotes-20121109_1_mike-tyson-undisputed-truth-famous-quotes.

[72] McGonigal, Kelly. *The Willpower Instinct*. New York: Penguin Group, 2012.

Mischel, Walter. *The Marshmallow Test*. Back Bay Books, 2015.

[73] McGonigal, Kelly. *The Willpower Instinct*. New York: Penguin Group, 2012.

[74] Lisle, Douglas. "4 Ways to Increase Your Willpower." *T. Colin Campbell Center for Nutrition Studies*, 12 Dec. 2014, nutrition-studies.org/4-ways-to-increase-your-willpower.

[75] Ratey, John J. *Spark: The Revolutionary New Science of Exercise and the Brain*. Little Brown, 2012.

[76] Ratey, John J. *Spark: The Revolutionary New Science of Exercise and the Brain*. Little Brown, 2012.

[77] Piepmeier, Aaron T. and Jennifer L. Etnier. "Brain-derived neurotrophic factor (BDNF) as a potential mechanism of the effects of acute exercise on cognitive performance." *Journal of Sport and Health Science*, Volume 4, Issue 1, Mar. 2015, www.sciencedirect.com/science/article/pii/S2095254614001161.

[78] "Turmeric." *drweil.com*. Accessed 17 Jul. 2015, www.drweil.com/drw/u/REM00019/Turmeric-Dr-Weils-Herbal-Remedies.html.

[79] Medina, John. *Brain Rules: 12 Principles for Surviving and Thriving at Work, Home, and School*. Seattle, WA: Pear, 2008.

[80] Brookshire, Bethany. "What's fuel for the body is fuel for the brain: a story of glycogen." *Scicurious*, 29 Feb. 2012, scicurious.scientopia.org/2012/02/29/whats-fuel-for-the-body-is-fuel-for-the-brain-a-story-of-glycogen.

[81] Nokia, Miriam S., Sanna Lensu, Juha P. Ahtiainen, Petra P. Johansson, Lauren G. Koch, Steven L. Britton, and Heikki Kainulainen. "Physical exercise increases adult hippocampal neurogenesis in male rats provided it is aerobic and sustained." *The Journal of Physiology*, 24 Feb. 2016, onlinelibrary.wiley.com/doi/10.1113/JP271552/abstract.

[82] Weinberg, Lisa , Anita Hasni, Minoru Shinohara, Audrey Duarte, "A single bout of resistance exercise can enhance episodic memory performance." Acta Psychologica, vol. 153, Nov. 2014, pp. 13-19, www.ncbi.nlm.nih.gov/pubmed/25262058.

[83] Small, Gary. "Can Exercise Cure Depression?" *Psychology Today*, 25 Sep. 2010, www.psychologytoday.com/blog/brain-boot-camp/201009/can-exercise-cure-depression.

[84] Medina, John. *Brain Rules: 12 Principles for Surviving and Thriving at Work, Home, and School.* Seattle, WA: Pear, 2008.

[85] "Walking Boosts Brain Function, Study Shows." *Live Science*, 30 Aug. 2010, www.livescience.com/34850-walking-boosts-brain-function-study-shows.html.

[86] Mitchell, Dan. "Silicon Valley's different kind of power walk." *Fortune*, 15 Nov. 2011, fortune.com/2011/11/15/silicon-valleys-different-kind-of-power-walk.

[87] Neithercott, Tracy. "A User's Guide to Insulin." *Diabetes Forecast*, April 2010, www.diabetesforecast.org/2010/apr/a-user-s-guide-to-insulin.html.

[88] Valls-Pedret, Cinta, Aleix Sala-Vila, Mercè Serra-Mir, Dolores Corella, Rafael de la Torre, Miguel Ángel Martínez-González, Elena H. Martínez-Lapiscina, Montserrat Fitó, Pérez-Heras, Jordi Salas-Salvadó, Ramon Estruch, Emilio Ros. "Mediterranean Diet and

Age-Related Cognitive Decline: A Randomized Clinical Trial." *JAMA Internal Medicine*, Jul. 2015, vol. 175, no. 7, jamanetwork.com/journals/jamainternalmedicine/fullarticle/2293082.

[89] Perlmutter, David, and Carol Colman. *The Better Brain Book: The Best Tools for Improving Memory, Sharpness, and Preventing Aging of the Brain.* New York: Riverhead, 2004.

[90] Reinberg, Steven. "Type 2 Diabetes May Shrink the Brain." *WebMD*, 29 Apr. 2014, www.webmd.com/diabetes/news/20140429/type-2-diabetes-may-shrink-the-brain-study-suggests.

[91] Blaszczak-Boxe, Agata. "Trans Fat May Impair Memory." *Live Science*, 17 Jun. 2015, www.livescience.com/51251-trans-fat-may-impair-memory.html.

[92] Schmidt, Elaine. "This is your brain on sugar: UCLA study shows high-fructose diet sabotages learning, memory." *UCLA Newsroom*, 15 May 2012, newsroom.ucla.edu/releases/this-is-your-brain-on-sugar-ucla-233992.

[93] "This Is Your Body on Soda," *Experience Life*, Apr. 2014, experiencelife.com/article/this-is-your-body-on-soda.

[94] "Added Sugars." *American Heart Association*, 20 Jul. 2016, www.heart.org/HEARTORG/HealthyLiving/HealthyEating/Nutrition/Added-Sugars_UCM_305858_Article.jsp#.V3UrNKsjHp4.

[95] "This Is Your Body on Soda," *Experience Life*, Apr. 2014, experiencelife.com/article/this-is-your-body-on-soda.

[96] "Do diet soft drinks actually make you gain weight?" *NHS Choices*, 8 Apr. 2015, www.nhs.uk/news/2015/04April/Pages/Do-diet-soft-drinks-actually-make-you-gain-weight.aspx.

[97] Doyle, Katharine. "Drinking diet soda linked to a widening waistline with age." *Reuters*, 18 Mar. 2015, www.reuters.com/article/us-health-aging-soda-belly-idUSKBN0ME2MH20150318.

[98] "The REAL story about alcohol and other drugs." *National Council on Alcoholism and Drug Dependence Inc*, 26 Jun. 2015, www.ncadd.org/about-addiction/underage-issues/the-real-story-about-alcohol-and-other-drugs.

[99] Healy, Melissa. "Sugary drinks linked to 25,000 deaths in the U.S. each year." *LA Times*, 29 Jun. 2015, www.latimes.com/science/sciencenow/la-sci-sn-sugary-soda-death-toll-20150629-story.html.

[100] "Alcohol – Low to Moderate." *CognitiveVitality.org*, 17 Jun. 2016, www.alzdiscovery.org/cognitive-vitality/report/low-to-moderate-alcohol-consumption.

[101] Paula, Elle. "Five Top Things to Drink to Be Healthy." *SFGATE*, healthyeating.sfgate.com/five-top-things-drink-healthy-7447.html.

[102] Cummings, David E. and Joost Overduin. "Gastrointestinal regulation of food intake." *The Journal of Clinical Investigation*, 2 Jan. 2007, vol. 117, no. 1, pp. 13-23, www.jci.org/articles/view/30227.

[103] "Neurological Researchers Find Fat May Be Linked to Memory Loss." *Rush University Medical Center*, 8 Oct. 2013, www.rush.edu/news/press-releases/neurological-researchers-find-fat-may-be-linked-memory-loss.

[104] Fiegl, Amanda. "A Brief History of Chocolate." *Smithsonian.com*, 1 Mar. 2008, www.smithsonianmag.com/arts-culture/a-brief-history-of-chocolate-21860917/?no-ist.

[105] "Heart Health Benefits of Chocolate." *Cleveland Clinic*, Jan. 2012, my.clevelandclinic.org/services/heart/prevention/nutrition/food-choices/benefits-of-chocolate.

[106] Heidi Godman, "Cocoa: a sweet treat for the brain?" *Harvard Health Blog*, 5 Feb. 2015, www.health.harvard.edu/blog/cocoa-sweet-treat-brain-201502057676.

[107] Heidi Godman, "Cocoa: a sweet treat for the brain?" Harvard Health Blog, 5 Feb. 2015, www.health.harvard.edu/blog/cocoa-sweet-treat-brain-201502057676.

[108] "Animals' Sleep: Is There A Human Connection?" *National Sleep Foundation*, sleepfoundation.org/sleep-news/animals-sleep-there-human-connection.

[109] "Survey: Americans know how to get better sleep – but don't act on it." *The Better Sleep Council,* bettersleep.org/better-sleep/the-science-of-sleep/sleep-statistics-research/better-sleep-survey.

[110] Institute of Medicine Committee on Sleep Medicine and Research; Colten HR, Altevogt BM, ed. "Sleep Disorders and Sleep Deprivation: An Unmet Public Health Problem." *National Academies Press*, 2006, www.ncbi.nlm.nih.gov/books/NBK19961.

[111] Payne, Jessica D. and Lynn Nadel. "Sleep, dreams, and memory consolidation: The role of the stress hormone cortisol." *Learn Memory*, Nov. 2004, vol. 11, no. 6, pp. 671–678, www.ncbi.nlm.nih.gov/pmc/articles/PMC534695.

[112] Anwar, Yasmin. "As we sleep, speedy brain waves boost our ability to learn." *Berkeley News*, 8 Mar. 2011, news.berkeley.edu/2011/03/08/sleep-brainwaves.

[113] Clear, James. "How to Get Better Sleep: The Beginner's Guide to Overcoming Sleep Deprivation." *JamesClear.com* (blog). jamesclear.com/better-sleep.

[114] "Who's at Risk?" *National Sleep Foundation*, drowsydriving.org/about/whos-at-risk.

[115] "Sleep Studies." *National Sleep Foundation*, sleepfoundation.org/sleep-topics/sleep-studies.

[116] "Talking Points About Roadway Users." *American Society of Civil Engineers*, www.asce.org/uploadedFiles/Technical_Areas/Transportation_and_Development_Engineering/Content_Pieces/put-the-brakes-on-fatalities-day-talking-points.pdf.

[117] Garnett, Carla. "Rest for the Weary? Shorter Work Shifts Suggested for Physicians." *NIH Record*, 5 Oct. 2007, vol. 59, no. 20, nihrecord.nih.gov/newsletters/2007/10_05_2007/story1.htm.

[118] Caba, Justin. "Sleep Experts Say Bosses Should Let Their Employees Take A Nap At Work To Boost Productivity." *Medical Daily*, 6 Jun. 2014, www.medicaldaily.com/sleep-experts-say-bosses-should-let-their-employees-take-nap-work-boost-productivity-286842.

[119] Mastin, Luke. "Sleep-Wake Homeostasis." *How Sleep Works*, 2013, www.howsleepworks.com/how_homeostasis.html.

[120] "Healthy Sleep Tips." *National Sleep Foundation*, sleepfoundation.org/excessivesleepiness/sleep-tools-tips/healthy-sleep-tips. (*1-4 in list)

[121] *Surrey Sleep Research Center*, University of Surrey, www.surrey.ac.uk/fhms/research/centres/ssrc. (*5-8 in list)

[122] Iliff, Jeffrey. "How our brains wash away the gunk during sleep." *Brain Institute,* 30 Oct. 2013, www.ohsu.edu/blogs/brain/2013/10/30/how-our-brains-wash-away-the-gunk-during-sleep.

"Brain's 'Garbage Truck' May Hold Key to Treating Alzheimer's and Other Disorders." *Neuroscience News,* 27 Jun. 2013, neuroscience-news.com/neurodegeneration-glymphatic-system-alzheimers-disease-269.

[123] Kirchgessner, A.L. and M.D. Gershon. "Innervation of the Pancreas by Neurons in the Gut." *The Journal of Neuroscience,* vol. 10, no. 5, May 1990, pp. 1626-1642, www.jneurosci.org/content/10/5/1626.full.pdf.

[124] Summa, Keith and Fred Turek. "The Clocks Within Us". Scientific American, vol. 312, no. 2, 20 Jan. 2015, www.nature.com/scientificamerican/journal/v312/n2/full/scientificamerican0215-50.html?WT.ec_id=SCIENTIFICAMERICAN-201502.

[125] Foer, Joshua and Michael Siffre. "Caveman: An Interview with Michel Siffre." *Cabinet,* no. 30, 2008, www.cabinetmagazine.org/issues/30/foer.php.

[126] Foer, Joshua and Michael Siffre. "Caveman: An Interview with Michel Siffre." *Cabinet,* no. 30, 2008, www.cabinetmagazine.org/issues/30/foer.php.

[127] Bear, Mark F., Barry W. Connors and Michael A. Paradiso. *Neuroscience: Exploring the Brain.* Baltimore: Williams and Wilkins, 1996.

Herzog, E.D., J.S. Takahashi and G.D. Block. "Clock controls circadian period in isolated suprachiasmatic nucleus neurons." *Nature Neuroscience,* vol. 1, no. 8, Dec. 1998, pp. 708-713, www.ncbi.nlm.nih.gov/

pubmed/10196587.

Lydic, Ralph, H. Elliott Albers, Beverly Tepper, Martin C. Moore-Ede.

"Three-dimensional structure of the mammalian suprachiasmatic nuclei: a comparative study of five species." *The Journal of Comparative Neurology*, vol. 204, no. 3, 20 Jan. 1982, pp. 225-237, onlinelibrary. wiley.com/wol1/doi/10.1002/cne.902040303/abstract.

Van den Pol, Anthony N. "The hypothalamic suprachiasmatic nucleus of rat: Intrinsic anatomy." *The Journal of Comparative Neurology*, vol. 191, no. 4, 15 Jun. 1980, 661-702, onlinelibrary.wiley.com/doi/10.1002/cne.901910410/full.

[128] National Sleep Foundation. "Sleep Drive and Your Body Clock." sleepfoundation.org/sleep-topics/sleep-drive-and-your-body-clock

[129] Clear, James. "How to Get Better Sleep: The Beginner's Guide to Overcoming Sleep Deprivation." *JamesClear.com* (blog), jamesclear.com/better-sleep.

[130] Clear, James. "How to Get Better Sleep: The Beginner's Guide to Overcoming Sleep Deprivation." *JamesClear.com* (blog), jamesclear.com/better-sleep.

[131] Medina, John. *Brain Rules: 12 Principles for Surviving and Thriving at Work, Home, and School.* Seattle, WA: Pear, 2008.

[132] American Sleep Association. "Jet Lag." Sep. 2007, www. sleepassociation.org/patients-general-public/jet-lag-2.

[133] National Sleep Foundation. "Sleep-wake cycle: Its Physiology and impact on Health." 2006, sleepfoundation.org/sites/default/files/SleepWakeCycle.pdf.

[134] National Sleep Foundation. "Sleep-wake cycle: Its Physiology and impact on Health." 2006, sleepfoundation.org/sites/default/files/SleepWakeCycle.pdf.

[135] "Powerful link between circadian rhythms and metabolism." *News-Medical.net*, 7 Apr. 2012, www.news-medical.net/news/20120407/Powerful-link-between-circadian-rhythms-and-metabolism.aspx.

[136] "Everything You Ever Wanted to Know About Drinking Coffee." *Conversational.com*, 17 Feb. 2016, www.conversational.com/everything-you-ever-wanted-to-know-about-drinking-coffee.

[137] Bradberry, Travis. "Caffeine: The Silent Killer of Success." *Forbes*, 21 Oct. 2012, www.forbes.com/sites/travisbradberry/2012/08/21/caffeine-the-silent-killer-of-emotional-intelligence/#46.

[138] Wenk, Gary Lee. *Your Brain on Food: How Chemicals Control Your Thoughts and Feelings.* Oxford: Oxford UP, 2010.

[139] Mah, Cheri D., Kenneth E. Mah, Eric J. Kezirian, William C. Dement. "The Effects of Sleep Extension on the Athletic Performance of Collegiate Basketball Players." *Sleep*, vol. 34, no. 7, 1 Jul. 2011, 943–950. www.ncbi.nlm.nih.gov/pmc/articles/PMC3119836.

[140] Merryman, Ashley. "Sleep Is For Wusses – An Idea Kids Are Learning All Too Well." *NurtureShock*, 10 Oct. 2007, www.pobronson.com/blog/2007/10/sleep-is-for-wusses-idea-kids-are.html.

[141] Spaeth, Andrea M., David F. Dinges, Namni Goel. "Managing neurobehavioral capability when social expediency trumps biological imperatives." *Progressive Brain Research,* vol. 199, 2012, pp. 377-398. www.ncbi.nlm.nih.gov/pubmed/22877676

[142] Mischel, Walter. *The Marshmallow Test*. Back Bay Books, 2015.

[143] Erskine, James. "Resistance can be Futile: Investigating Behavioral Rebound." *Appetite,* vol. 50, no. 2-3, Mar. 2008, pp 415-421. www.researchgate.net/publication/5841201_Resistance_can_be_futile_Investigating_behavioural_rebound.

Erskine, James, and G.J. Georgiou, "Effects of Thought Suppression on Eating Behavior in Restrained and Non-Restrained Eaters." *Appetite,* vol. 54, no. 3, Feb. 2010, pp. 499-503. www.researchgate.net/publication/41427278_Effects_of_thought_suppression_on_eating_behavior_in_restrained_and_non-restrained_eaters.

[144] McGonigal, Kelly. *The Willpower Instinct*. New York: Penguin Group, 2012.

[145] McGonigal, Kelly. *The Willpower Instinct*. New York: Penguin Group, 2012.

[146] McGonigal, Kelly. *The Willpower Instinct*. New York: Penguin Group, 2012.

[147] Baumeister, Roy F., and John Tierney. *Willpower: Rediscovering the Greatest Human Strength*. New York: Penguin, 2011.

[148] http://well.blogs.nytimes.com/2012/02/22/how-exercise-fuels-the-brain/?_r=0

Chapter Four

[149] Levitin, Daniel J. *The Organized Mind: Thinking Straight in the Age of Information Overload*. London: Viking, 2014.

[150] Gerstner, Wulfram, Werner M. Kistler, Richard Naud and Liam Paninski. "19.1 Hebb Rule and Experiments." *Neuronal Dynamics: From single neurons to networks and models of cognition,* Cambridge University Press, 8 Sep. 2014, neuronaldynamics.epfl.ch/online/Ch19.S1.html

[151] "What is Your Reaction Time?" *Stanford University—Tech Museum of Innovation,* 2007, virtuallabs.stanford.edu/tech/images/ReactionTime.SU-Tech.pdf.

[152] Breindl, Anette. "Myelination Exhibits Plasticity, Links to Behavior in Adult Brain." *BioWorld,* www.bioworld.com/content/myelination-exhibits-plasticity-links-behavior-adult-brain-0.

[153] Bergland, Christopher. "Why Is Physical Activity So Good for Your Brain?" *Psychology Today,* 22 Sep. 2014, www.psychologytoday.com/blog/the-athletes-way/201409/why-is-physical-activity-so-good-your-brain.

[154] Dean, Jeremy. *Making Habits, Breaking Habits: Why We Do Things, Why We Don't, and How to Make Any Change Stick.* Boston: Da Capo Press, 2013.

[155] Coyle, Daniel. *The Talent Code: Greatness Isn't Born: It's Grown, Here's How.* New York: Bantam, 2009.

[156] Ericsson, Anders K., Ralf Th. Krampe, and Clemens Tesch-Romer. "The Role of Deliberate Practice in the Acquisition of Expert Performance." *Psychological Review,* vol. 100, no. 3, 1993, pp. 363-406, http://projects.ict.usc.edu/itw/gel/EricssonDeliberatePractice-PR93.PDF.

[157] Staal, Mark A., Amy E. Bolton, Rita A. Yaroush, and Lyle E. Borne, Jr. "Cognitive Performance and Resilience to Stress." *Biobehavioral Resilience to Stress,* 2008, psych.colorado.edu/~lbourne/CR.doc.

[158] Dunning, David, Chip Heath, and Jerry M. Suls. *Flawed Self-assessment: Implications for Health, Education, and the Workplace.* Malden, MA: Blackwell Pub., 2004.

[159] Rozenblit, Leonid, and Frank Keil. "The misunderstood limits of folk science: An illusion of explanatory depth." *Cognitive Science*, vol. 26, 2002, pp. 521-562. www.psy.cmu.edu/~siegler/Roz-Keil02.pdf.

[160] Isaacson, Melissa. "U.S. men's gymnastics team must let go of its London letdown." *ESPN.com*, 24 Jun. 2016, espn.go.com/olympics/gymnastics/story/_/id/16488446/olympic-trials-2016-us-men-gymnastics-team-let-go-london-letdown.

[161] Gladwell, Malcolm. *Outliers: The Story of Success.* New York: Little, Brown, 2008.

[162] Ericsson, Karl Anders, and Robert Pool. *Peak: Secrets from the New Science of Expertise.* London: Bodley Head, 2016.

[163] Ericsson, Karl Anders, and Robert Pool. *Peak: Secrets from the New Science of Expertise.* London: Bodley Head, 2016.

[164] Ericsson, Karl Anders, and Robert Pool. *Peak: Secrets from the New Science of Expertise.* London: Bodley Head, 2016.

[165] Ericsson, Karl Anders, and Robert Pool. *Peak: Secrets from the New Science of Expertise.* London: Bodley Head, 2016.

[166] Ericsson, Karl Anders, and Robert Pool. *Peak: Secrets from the New Science of Expertise.* London: Bodley Head, 2016.

[167] Lewisohn, Mark. *Tune in.* London: Little, Brown, 2013.

[168] "INTERVIEW: Paul McCartney heads to Canada." *CBC News*, 6 Aug. 2010, www.cbc.ca/news/arts/interview-paul-mccartney-heads-to-canada-1.942764.

[169] Simon, Herbert A., and William G. Chase. "Skill in chess." *American Scientist*, vol. 61, no. 4, Jul 1973, 394-403.

[170] Gerstner, Wulfram, Werner M. Kistler, Richard Naud and Liam Paninski. "19.1 Hebb Rule and Experiments." *Neuronal Dynamics: From single neurons to networks and models of cognition*, Cambridge University Press, 8 Sep. 2014, neuronaldynamics.epfl.ch/online/Ch19.S1.html.

[171] Imai, Masaaki. *Kaizen: (Ky'zen): The Key to Japan's Competitive Success*. New York: McGraw-Hill, 1986.

[172] "Dr W. Edwards Deming's 14 Principles - in full." *Qualityregister.co.uk*, www.qualityregister.co.uk/14principles.html.

[173] Davis, Rebecca. "The Doctor Who Championed Hand-Washing And Briefly Saved Lives." *NPR Morning Edition*, 12 Jan. 2015, www.npr.org/sections/health-shots/2015/01/12/375663920/the-doctor-who-championed-hand-washing-and-saved-women-s-lives.

[174] Ericsson, Anders. Personal interview. 3 Aug. 2016

[175] Schultz, Colin. "Just Twenty-Nine Dominoes Could Knock Down the Empire State Building." *Smithsonian.com*, 17 Jan. 2013, www.smithsonianmag.com/smart-news/just-twenty-nine-dominoes-could-knock-down-the-empire-state-building-2232941.

[176] Leerhsen, Charles. *Ty Cobb: A Terrible Beauty*. New York: Simon & Schuster, 2015.

[177] Kanfer, Stefan. *Groucho: The Life and times of Julius Henry Marx*. New York: Knopf, 2000.

[178] Kanfer, Stefan. *Groucho: The Life and times of Julius Henry Marx*. New York: Knopf, 2000.

[179] Vickers, Rebecca. *The Story Behind Mark Twain's The Adventures of Huckleberry Finn*. Chicago: Heineman Library, 2007.

[180] *Respectfully Quoted: A Dictionary of Quotations*. Edited by Suzy Platt, New York: Barnes & Noble Books, 1993.

[181] Twain, Mark. *The Innocents Abroad: Or, The New Pilgrims' Progress: Being Some Account of the Steamship Quaker City's Pleasure Excursion to Europe and the Holy Land*. New York: Harper & Bros., 1911.

[182] Ward, Geoffrey C., Ken Burns, and Dayton Duncan. *Mark Twain: An Illustrated Biography*. New York: Alfred A Knopf, 2001.

[183] Kaplan, Justin. *Mr. Clemens and Mark Twain: A Biography*. New York: Simon and Schuster,1966, pp. 180.

[184] Lorenzi, Rossella. "Michelangelo's marble may be flawed." *ABC Science*, 13 Sep. 2005, www.abc.net.au/science/articles/2005/09/13/1459036.htm.

[185] Ericsson, Karl Anders., and Robert Pool. *Peak: Secrets from the New Science of Expertise*. London: Bodley Head, 2016.

[186] Ericsson, Anders. Personal interview. 3 Aug. 2016.

[187] "July 20, 1969: One Giant Leap For Mankind." *NASA.gov*, 14 Jul. 2014, www.nasa.gov/mission_pages/apollo/apollo11.html.

[188] Espinoza, Erika. "What if Michael Phelps was his own country?" *Chicago Tribune,* 17 Aug. 2016, www.chicagotribune.com/sports/international/ct-phelps-medals-in-context-htmlstory.html.

[189] Futterman, Matthew. "Michael Phelps's Coach Shares His Secrets." *The Wall Street Journal,* 12 May 2016, www.wsj.com/articles/michael-phelpss-coach-shares-his-secrets-1463089497.

[190] "Michael Phelps on Making Olympic History." *CBS News,* 25 Nov. 2008, www.cbsnews.com/news/michael-phelps-on-making-olympic-history.

[191] "Interview With Guitar Tech Extraordinaire Rene Martinez." *Iconic Axes: The Instruments Used by the Gods of Six Strings.* 18 Sep. 2012, iconicaxes.blogspot.com/2012/09/interview-with-guitar-tech.html?view=magazine.

[192] Ford, Richard. "Richard Ford Reviews Bruce Springsteen's Memoir." *The New York Times,* 22 Sep. 2016, www.nytimes.com/2016/09/25/books/review/bruce-springsteen-born-to-run-richard-ford.html?emc=eta1.

[193] Springsteen, Bruce. *Born to Run.* New York: Simon & Schuster, 2016.

[194] Gardner, Howard. *Changing Minds: The Art and Science of Changing Our Own and Other People's Minds.* Boston, MA: Harvard Business School, 2006.

[195] Gardner, Howard. *Changing Minds: The Art and Science of Changing Our Own and Other People's Minds.* Boston, MA: Harvard Business School, 2006.

[196] Leonard, Michael. "Robert Johnson: 'King of the Delta Blues.'" *Gibson.com,* 19 Oct. 2012, www.gibson.com/News-Lifestyle/

Features/en-us/robert-johnson-king-of-the-delta-blues-101-2012.
aspx.

[197] Charters, Samuel. *The Country Blues*. New York: Da Capo, 1975.

Leonard, Michael. "Robert Johnson: 'King of the Delta Blues.'" *Gibson.com*, 19 Oct. 2012, www.gibson.com/News-Lifestyle/Features/en-us/robert-johnson-king-of-the-delta-blues-101-2012.aspx.

[198] "Biography." *Robert Johnson Blues Foundation*, www.robertjohnsonbluesfoundation.org/biography.

[199] "Biography." *Robert Johnson Blues Foundation*, www.robertjohnsonbluesfoundation.org/biography.

[200] Buncombe, Andrew. "The grandfather of rock'n'roll: The devil's instrument." *Independent*, 25 Jul. 2006, www.independent.co.uk/news/world/americas/the-grandfather-of-rocknroll-the-devils-instrument-409317.html.

[201] "Biography." Robert Johnson Blues Foundation, www.robertjohnsonbluesfoundation.org/biography.

[202] Polgar, Susan. Personal interview. 8 May 2015.

[203] Maass, Peter. "Home-grown Grandmasters." *The Washington Post*, 11 Mar. 1992, www.washingtonpost.com/archive/lifestyle/1992/03/11/home-grown-grandmasters/82928ac3-a934-4d0d-929b-d89ba020fa25.

[204] Polgár, László. *Nevelj zsenit! (Bring Up Genius)*, Budapest: Interart,1989.

[205] "Master Honorees." *International Swimming Hall of Fame*, 2010, www.ishof.org/tamás-széchy.html.

[206] László Polgár. *Chess: 5334 Problems, Combinations and Games*. Black Dog & Leventhal, 2013.

[207] Myers, Linnet. "Trained To Be A Genius, Girl, 16, Wallops Chess Champ Spassky For $110,000." *Chicago Tribune*, 18 Feb. 1993, articles.chicagotribune.com/1993-02-18/news/9303181339_1_judit-polgar-boris-spassky-world-chess-champion.

[208] Allott, Serena, "Queen takes all." *Telegraph*, 16 Jan. 2002, www.telegraph.co.uk/news/health/3294892/Queen-takes-all.html.

[209] "Chess Team Accomplishments." Webster University, www.webster.edu/spice/chess-team/accomplishments.html.

[210] Polgár, László. *Nevelj zsenit! (Bring Up Genius)*, Budapest: Interart,1989.

[211] Tobak, Steve. "Making a Team Greater Than the Sum of Its Parts." *CBS Moneywatch*, 26 Jul. 2011, www.cbsnews.com/news/making-a-team-greater-than-the-sum-of-its-parts.

[212] Allott, Serena, "Queen takes all." *Telegraph*, 16 Jan. 2002, www.telegraph.co.uk/news/health/3294892/Queen-takes-all.html.

[213] "A Conversation with Andre Agassi." *Fora.TV*, 20 Nov. 2009, library.fora.tv/2009/11/20/A_Conversation_with_Andre_Agassi.

[214] Zi, Sun. *The Art of War: Sun Zi's Military Methods*. Translated by Victor H. Mair, New York: Columbia UP, 2007.

[215] Flora, Carlin. "The Grandmaster Experiment." *Psychology Today*, 1 Jul. 2005, www.psychologytoday.com/articles/200507/the-grandmaster-experiment

[216] "Nona Gaprindashvili – the legendary fifth world champion." *Chess News*, 3 May 2006, en.chessbase.com/post/nona-gaprindashvili-the-legendary-fifth-world-champion.

[217] Bor, Daniel. *The Ravenous Brain: How the New Science of Consciousness Explains Our Insatiable Search for Meaning*. New York: Basic, 2012.

[218] "Episode 1: Make Me a Genius." *My Brilliant Brain*, National Geographic, episode 1, 2010, *YouTube*, uploaded 29 Apr. 2012, www.youtube.com/watch?v=2wzs33wvr9E.

[219] Grady, Denise. "The Vision Thing: Mainly in the Brain." *Discover*, 1 Jun. 1993, discovermagazine.com/1993/jun/thevision-thingma227.

[220] Bradt, Steve. "'Face-blindness' disorder may not be so rare." 1 Jun. 2006, *Harvard Gazette*, news.harvard.edu/gazette/story/2006/06/face-blindness-disorder-may-not-be-so-rare.

[221] Keefe, Patrick Radden. "The Detectives Who Never Forget a Face." *New Yorker*, 22 Aug. 2016, www.newyorker.com/magazine/2016/08/22/londons-super-recognizer-police-force.

[222] Gauthier, Isabel, Michael J. Tarr, Jill Moylan, Adam W. Anderson, Pawel Skudlarski, and John C. Gore. "Does Visual Subordinate-Level Categorisation Engage The Functionally Defined Fusiform Face Area?" *Cognitive Neuropsychology,* vol. 17, no. 1-3, 1 Feb. 2000, pp. 143-64. www.tandfonline.com/doi/abs/10.1080/026432900380544.

Tarr, Michael J., and Isabel Gauthier. "FFA: a flexible fusiform area for subordinate-level visual processing automatized by expertise." *Nature Neuroscience,* vol. 3, 2000, pp. 764 – 769, doi:10.1038/77666

223 "Episode 1: Make Me a Genius." *My Brilliant Brain*, National Geographic, episode 1, 2010, *YouTube*, uploaded 29 Apr. 2012, www.youtube.com/watch?v=2wzs33wvr9E.

224 "Did you know…?" *Chess-poster.com*, 2000, www.chess-poster.com/english/notes_and_facts/did_you_know.htm.

Chapter Five

225 Ericsson, Anders. "Superior Working Memory in Experts." *Florida University.*

226 Rock, David. New Study Shows Humans Are on Autopilot Nearly Half the Time." *Psychology Today*, 14 Nov. 2010, www.psychologytoday.com/blog/your-brain-work/201011/new-study-shows-humans-are-autopilot-nearly-half-the-time.

227 Dean, Jeremy. *Making Habits, Breaking Habits: Why We Do Things, Why We Don't, and How to Make Any Change Stick.* Boston: Da Capo Press, 2013.

228 Hinckley, David. "Average American watches 5 hours of TV per day, report shows." New York Daily News, 5 Mar. 2014, www.nydailynews.com/life-style/average-american-watches-5-hours-tv-day-article-1.1711954.

229 Wallace, David Foster. *This Is Water: Some Thoughts, Delivered on a Significant Occasion, about Living a Compassionate Life.* Little, Brown, and Company, 2009.

[230] Duhigg, Charles. "How Companies Learn Your Secrets. *New York Times*, 16 Feb. 2012.

[231] Duhigg, Charles. "How Companies Learn Your Secrets. *New York Times*, 16 Feb. 2012.

[232] Duhigg, Charles. "How Companies Learn Your Secrets. *New York Times*, 16 Feb. 2012.

[233] Duhigg, Charles. *The Power of Habit: Why We Do What We Do in Life and Business.* New York: Random House, 2012.

[234] Dean, Jeremy. *Making Habits, Breaking Habits: Why We Do Things, Why We Don't, and How to Make Any Change Stick.*

[235] Maltz, Maxwell. *Psycho-cybernetics: A New Way to Get More Living out of Life.* Englewood Cliffs, NJ: Prentice-Hall, 1960.

[236] Maltz, Maxwell. *Psycho-Cybernetics: Updated and Expanded.* New York: Perigee, 2015.

[237] Lally, Phillippa, Cornelia H. M. van Jaarsveld, Henry W. W. Potts, and Jane Wardle. "How Are Habits Formed: Modeling Habit Formation in the Real World." *European Journal of Social Psychology,* 16 Jul. 2009, doi: 10.1002/ejsp.674.

[238] Frijda, Nico H. "The Law of Emotion." *American Psychologist*, vol. 40, no. 5, May 1988, pp. 349-358. dx.doi.org/10.1037/0003-066X.43.5.349.

[239] McLeod, Saul. "Pavlov's Dogs." *Simply Psychology,* 2007, www.simplypsychology.org/pavlov.html.

[240] Bailey, Regina. "Corpus Callosum." *About.com,* biology. about.com/od/anatomy/p/corpus-callosum.htm.

[241] Thorn, Catherine A., Hisham Atallah, Mark Howe, and Ann M. Graybiel. "Differential dynamics of activity changes in dorsolateral and dorsomedial striatal loops during learning." *Neuron,* vol. 66, no. 5, 2010, pp. 781-795. doi: 10.1016/j.neuron.2010.04.036.

Ashby, F. Gregory and John M. Ennis, "The Role of the Basal Ganglia in Category Learning," *Psychology of Learning and Motivation,* vol. 46, 2006, pp. 1-36, labs.psych.ucsb.edu/ashby/gregory/reprints/ BasalGangliaSS.pdf.

Duhigg, Charles. *The Power of Habit: Why We Do What We Do in Life and Business.* New York: Random House, 2012.

[242] Rack, Jessie. "Are You Flossing Or Just Lying About Flossing? The Dentist Knows." *NPR.org,* 24 Jun. 2015, www.npr.org/sections/health-shots/2015/06/24/417184367/are-you-flossing-or-just-lying-about-flossing-the-dentist-knows.

[243] Suckling, Lee. "How effective is flossing?" *Stuff.co.nz,* 1 Sep 2016, www.stuff.co.nz/life-style/well-good/teach-me/83710429/how-effective-is-flossing.

Saint Louis, Catherine. "Feeling Guilty About Not Flossing? Maybe There's No Need." *New York Times,* 2 Aug. 2016, www.nytimes.com/2016/08/03/health/flossing-teeth-cavities.html?_r=0;

[244] Corliss, Julie. "Treating gum disease may lessen the burden of heart disease, diabetes, other conditions." *Harvard Health Blog,* 23 Jul. 2014, www.health.harvard.edu/blog/treating-gum-disease-may-lessen-burden-heart-disease-diabetes-conditions-201407237293.

245 Reblin, Maija, and Bert N. Uchino. "Social and emotional support and its implication for health." *Current Opinion in Psychiatry*, vol. 21, no. 2, Apr. 2008, pp. 201-205, www.researchgate.net/publication/5518916_Social_and_emotional_support_and_its_implication_for_health.

246 McRaney, David. "Extinction Burst." *You Are Not So Smart*, 7 Jul. 2010, youarenotsosmart.com/2010/07/07/extinction-burst.

247 Homer, and Richmond Lattimore. *The Odyssey of Homer*. New York: Harper & Row, 1967.

248 Fogg, BJ. "A Behavior Model for Persuasive Design." *Bjfogg.com*, bjfogg.com/fbm_files/page4_1.pdf.

249 Fogg, BJ. "A Behavior Model for Persuasive Design." *Bjfogg.com*, bjfogg.com/fbm_files/page4_1.pdf.

250 Zelman, Kathleen M. "The Olympic Diet of Michael Phelps." *WebMD*, 13 Aug. 2008, www.webmd.com/diet/20080813/the-olympic-diet-of-michael-phelps.

251 "Isaac Newton Biography." *Biography.com*, www.biography.com/people/isaac-newton-9422656#professional-life.

252 Fogg, BJ. "A Behavior Model for Persuasive Design." *Bjfogg.com*, bjfogg.com/fbm_files/page4_1.pdf.

253 "Bottled Water Facts." www.banthebottle.net/bottled-water-facts.

254 Noted that Elkay Corporation is one of my employers, but I entered into no agreement, and was not asked to write about, promote, or was specifically compensated for this story

[255] Damasio, Antonio R. *Self Comes to Mind: Constructing the Conscious Brain*. New York: Pantheon, 2010.

[256] Damasio, Antonio R. *Self Comes to Mind: Constructing the Conscious Brain*. New York: Pantheon, 2010.

[257] Fogg, BJ. "Broken routines, broken habits (that's normal)." *Sandbox for Tiny Habits W/ BJ Fogg*, 22 Oct. 2015, tinyhabits.com/sandbox.

[258] Oettingen, Gabriele. *Rethinking Positive Thinking: Inside the New Science of Motivation*. New York: Current, 2015.

[259] Higdon, Hal. *Marathon: The Ultimate Training Guide*. Emmaus, Penn.: Rodale, 2005.

[260] Douglas, Scott. "2017 Boston Marathon Registration FAQ." *Runners World*, 21 Apr. 2016, www.runnersworld.com/boston-marathon/2017-boston-marathon-registration-faq.

[261] Gollwitzer, Peter M. and Paschal Sheeran. "Implementation Intentions and Goal Achievement: A Meta-Analysis of Effects and Processes." *Advances in Experimental Psychology*, vol. 38, 2006, pp. 69-119. doi: 10.1016/S0065-2601(06)38002-1.

[262] Achziger, Anja, Peter M. Gollwitzer, and Paschal Sheeran. "Implementation Intentions and Shielding Goal Striving From Unwanted Thoughts and Feelings." *Personality and Social Psychology Bulletin*, vol. 34, 2008, pp. 381. doi: 10.1177/0146167207311201.

[263] Achziger, Anja, Peter M. Gollwitzer, and Paschal Sheeran. "Implementation Intentions and Shielding Goal Striving From Unwanted Thoughts and Feelings." *Personality and Social Psychology Bulletin*, vol. 34, 2008, pp. 381. doi: 10.1177/0146167207311201.

264 Clear, James. "How to Stop Procrastinating on Your Goals by Using the "Seinfeld Strategy." *James Clear* (blog), jamesclear.com/stop-procrastinating-seinfeld-strategy.

265 _Seinfeld [Jerry Seinfeld]. "Jerry Seinfeld here. I will give you an answer." Reddit, 6 Jan. 2014, www.reddit.com/r/IAmA/comments/1ujvrg/jerry_seinfeld_here_i_will_give_you_an_answer.

266 Safire, William. "ON LANGUAGE; Ockham's Razor's Close Shave." The New York Times Magazine, 31 Jan. 1999, www.nytimes.com/1999/01/31/magazine/on-language-ockham-s-razor-s-close-shave.html.

267 Vandekerckhove, Joachim, Dora Matzke, Eric-Jan Wagenmakers. "Model Comparison and the Principle of Parsimony." *The Oxford Handbook of Computational and Mathematical Psychology*, edited by Jerome R. Busemeyer, Zheng Wang, James T. Townsend, and Ami Eidels, Oxford University, 2014, pp. 300-318. www.cidlab.com/prints/vandekerckhove2014model.pdf.

Chapter 6

268 "The Nobel Prize in Physiology or Medicine 1981". *Nobelprize.org*, Nobel Media AB 2014, www.nobelprize.org/nobel_prizes/medicine/laureates/1981.

269 Brogaard, Berit. "Split Brains." *Psychology Today*, 6 Nov. 2012, www.psychologytoday.com/blog/the-superhuman-mind/201211/split-brains.

270 "What is Lateralization?" *Allpsychologycareers.com*, QuinStreet Inc., www.allpsychologycareers.com/topics/lateralization-right-brain-left-brain.html.

271 Clancy, Kelly. "The Strangers in Your Brain." *New Yorker Magazine*, 17 Oct. 2015, www.newyorker.com/tech/elements/the-strangers-in-your-brain.

Hawkins, Jeff. "How Brain Science Will Change Computing." *www.ted.com*, Feb. 2003, www.ted.com/talks/jeff_hawkins_on_how_brain_science_will_change_computing?language=en.

272 Becker, Kirk A. "History of the Stanford-Binet Intelligence Scales: Content and Psychometrics." *Stanford-Binet Intelligence Scales, Fifth Edition, Assessment Service Bulletin Number 1,* Riverside Publishing, 2003, www.hmhco.com/~/media/sites/home/hmh-assessments/clinical/stanford-binet/pdf/sb5_asb_1.pdf?la=en.

273 Andreasen, Nancy C. *The Creating Brain: The Neuroscience of Genius*. New York: Dana, 2005.

274 Andreasen, Nancy C. "Secrets of the creative brain." *The Atlantic*, Jul. 2014, www.theatlantic.com/magazine/archive/2014/07/secrets-of-the-creative-brain/372299.

275 Holahan, C. K., and Sears, R. R. *The Gifted Group in Later Maturity*. Stanford, CA: Stanford University Press, 1995.

276 Terman, Lewis M. *Genetic Studies of Genius*. Stanford, CA: Stanford University Press, 1926.

277 Gladwell, Malcolm. *Outliers*. New York: Little, Brown and Company, 2008. pp. 90.

278 "Luis Alvarez - Biographical". *Nobelprize.org*. Nobel Media AB 2014. www.nobelprize.org/nobel_prizes/physics/laureates/1968/alvarez-bio.html.

[279] "The Nobel Prize in Physics 1956". *Nobelprize.org.* Nobel Media AB 2014. www.nobelprize.org/nobel_prizes/physics/laureates/1956.

[280] Saxon, Wolfgang. "William B. Shockley, 79, Creator of Transistor and Theory on Race." *The New York Times,* 14 Aug. 1989, www.nytimes.com/learning/general/onthisday/bday/0213.html.

[281] Saxon, Wolfgang. "William B. Shockley, 79, Creator of Transistor and Theory on Race." The New York Times, 14 Aug. 1989, www.nytimes.com/learning/general/onthisday/bday/0213.html.

[282] (Shockley died in 1989, his legacy tarnished by his support of "retrogressive evolution", a concept closely associated with Eugenics. It should be noted that Lewis M. Terman, like Shockley, was also a believer in Eugenics. This serves as a sobering reminder that advanced thinking in one domain—physics—does not equate to wisdom in unrelated domains like social science, ethics, or biology.)

[283] Tajnai, Carolyn. "FRED TERMAN, THE FATHER OF SILICON VALLEY." *Stanford Computer Forum*, Stanford University, May 1985, forum.stanford.edu/carolyn/terman.

[284] Terman, Lewis Madison; M. H. Oden. *The Gifted Child Grows Up: Twenty-five Years' Follow-up of a Superior Group. (Genetic Studies of Genius, Vol. 4,* 4th ed., Stanford University Press, 1947. p. 352.

[285] Kaufman, Scott Barry. "Learning Strategies Outperform IQ in Predicting Achievement." *Scientific American*, 8 Apr. 2013, blogs.scientificamerican.com/beautiful-minds/learning-strategies-outperform-iq-in-predicting-achievement.

[286] Andreasen, Nancy C. "Secrets of the creative brain." *The Atlantic*, Jul. 2014, www.theatlantic.com/magazine/archive/2014/07/secrets-of-the-creative-brain/372299.

[287] Gardner, Howard. *Frames of Mind: The Theory of Multiple Intelligences*. New York: Basic, 1983.

[288] Golis, Chris. "A Brief History of Emotional Intelligence." *Practical Emotional Intelligence*, 2013, www.emotionalintelligence-course.com/eq-history.

[289] Goleman, Daniel, and Daniel Goleman. *Emotional Intelligence: Why It Can Matter More than IQ: & Working with Emotional Intelligence*. London: Bloomsbury, 2004.

[290] Anderson, Todd. "Test Your Creativity: 5 Classic Creative Challenges." *99U*, 2012, 99u.com/articles/7160/test-your-creativity-5-classic-creative-challenges.

[291] Baer, John. "Domain specificity of creativity: Theory, research, and practice." *TEXT Special Issue: Creativity: Cognitive, Social and Cultural Perspectives*, edited by Nigel McLoughlin & Donna Lee Brien, Apr. 2012, www.textjournal.com.au/speciss/issue13/Baer.pdf.

[292] Andreasen, Nancy C. "Secrets of the creative brain." *The Atlantic*, Jul. 2014, www.theatlantic.com/magazine/archive/2014/07/secrets-of-the-creative-brain/372299.

[293] Dietrich, Arne. *How Creativity Happens in the Brain*. Palgrave Macmillan, 2015.

[294] Alfred, Randy. "Oct. 21, 1879: Edison Gets the Bright Light Right." *Wired.com*, 21 Oct. 2009, www.wired.com/2009/10/1021edison-light-bulb.

[295] Irving, John. *A Prayer for Owen Meany: A Novel*. New York: Morrow, 1989.

[296] Dietrich, Arne. *How Creativity Happens in the Brain*. Palgrave Macmillan, 2015.

[297] Mayo Clinic Staff, "C Difficile Infection." 18 Jun. 2016, *Mayoclinic.org*, www.mayoclinic.org/diseases-conditions/c-difficile/home/ovc-20202264.

[298] Dietrich, Arne. *How Creativity Happens in the Brain*. Palgrave Macmillan, 2015.

[299] Biello, David. "Fact or Fiction?: Archimedes Coined the Term "Eureka!" in the Bath." *Scientific American*, 8 Dec. 2006, www.scientificamerican.com/article/fact-or-fiction-archimede.

[300] Migliore, Lauren. "The Aha Moment." *BrainWorld*, 14 Jun. 2012, brainworldmagazine.com/the-aha-moment/#sthash.oaUZSjt2.dpuf.

[301] Raudenbush, David. "Critical Thinking Brain Teasers." *Study.com*, study.com/academy/lesson/critical-thinking-brain-teasers.html.

[302] "Epiphany." *Oxforddictionaries.com*, Oxford University Press, www.oxforddictionaries.com/us/definition/american_english/epiphany.

[303] Weinschenk, Susan. "Our Minds Wander at Least 30 Percent of the Time." *Psychology Today*, 10 Jan. 2013, www.psychologytoday.com/blog/brain-wise/201301/our-minds-wander-least-30-percent-the-time.

[304] Weinschenk, Susan. *How to Get People to Do Stuff: Master the Art and Science of Persuasion and Motivation.* Berkeley, CA: New Riders, 2013.

[305] Weinschenk, Susan. "Our Minds Wander at Least 30 Percent of the Time." *Psychology Today*, 10 Jan. 2013, www.psychology-today.com/blog/brain-wise/201301/our-minds-wander-least-30-percent-the-time.

[306] Bernstein, Aaron. *Naturwissenschaftliche Volksbücher.* Berlin: Duncker, 1874.

[307] Isaacson, Walter. *Einstein: His Life and Universe.* Simon & Schuster, 2008.

[308] "MIT Research - Brain Processing of Visual Information." MIT News, 19 Dec. 1996, news.mit.edu/1996/visualprocessing.

[309] Isaacson, Walter. *Einstein: His Life and Universe.* Simon & Schuster, 2008.

[310] Popova, Maria. "How Einstein Thought: Why "Combinatory Play" Is the Secret of Genius" *Brainpickings.org*, 14 Aug. 2013, www.brainpickings.org/2013/08/14/how-einstein-thought-combinatorial-creativity.

[311] Dietrich, Arne. Personal interview. 22 Jul. 2016.

[312] Mastin, Luke. "17TH CENTURY MATHEMATICS – NEWTON." Storyofmathematics.com, 2010. www.storyofmathematics.com/17th_newton.html.

Andreasen, Nancy C. "Secrets of the creative brain." *The Atlantic*, Jul. 2014, www.theatlantic.com/magazine/archive/2014/07/secrets-of-the-creative-brain/372299.

[313] Andreasen, Nancy C. "Secrets of the creative brain." *The Atlantic*, Jul. 2014, www.theatlantic.com/magazine/archive/2014/07/secrets-of-the-creative-brain/372299.

[314] "Prince Bio." *Rollingstone.com,* www.rollingstone.com/music/artists/prince/biography.

[315] Ferguson, Kirby. "Everything is a Remix Part 3". *Vimeo*, 20 June 2011, vimeo.com/25380454.

Ferguson, Kirby. "Everything is a Remix Part 1". *Vimeo*, 12 Sept. 2010, vimeo.com/84954874.

Ferguson, Kirby. "Everything is a Remix Part 2". *Vimeo*, 1 Feb. 2011, vimeo.com/19447662.

[316] Ferguson, Kirby. "Everything is a Remix Part 1". *Vimeo*, 12 Sept. 2010, vimeo.com/84954874.

Ferguson, Kirby. "Everything is a Remix Part 2". *Vimeo*, 1 Feb. 2011, vimeo.com/19447662.

Ferguson, Kirby. "Everything is a Remix Part 3". *Vimeo*, 20 June 2011, vimeo.com/ 25380454.

[317] Ferguson, Kirby. "Everything is a Remix Part 3". *Vimeo*, 20 June 2011, vimeo.com/ 25380454.

Ferguson, Kirby. "Everything is a Remix Part 1". *Vimeo*, 12 Sept. 2010, vimeo.com/84954874.

[318] Ferguson, Kirby. "Everything is a Remix Part 1". *Vimeo*, 12 Sept. 2010, vimeo.com/84954874.

Ferguson, Kirby. "Everything is a Remix Part 2". *Vimeo*, 1 Feb. 2011, vimeo.com/19447662.

Ferguson, Kirby. "Everything is a Remix Part 3". *Vimeo*, 20 June 2011, vimeo.com/ 25380454.

[319] Ferguson, Kirby. "Everything is a Remix Part 2". *Vimeo*, 1 Feb. 2011, vimeo.com/19447662.

Ferguson, Kirby. "Everything is a Remix Part 3". *Vimeo*, 20 June 2011, vimeo.com/ 25380454.

Ferguson, Kirby. "Everything is a Remix Part 1". *Vimeo*, 12 Sept. 2010, vimeo.com/84954874.

[320] Rakhlin, Serge. "Tarkovsky And The Revenant – Homage, And Beyond." *Goldenglobes.com*, 18 Feb. 2016, www.goldenglobes. com/articles/tarkovsky-and-revenant---homage-and-beyond.

[321] Hiatt, Brian. "'Hamilton': Meet the Man Behind Broadway's Hip-Hop Masterpiece." *Rolling Stone*, 29 Sept. 2015, www.rolling-stone.com/culture/features/hamilton-meet-the-man-behind-broad-ways-hip-hop-masterpiece-20150929.

[322] "Alexander Hamilton > Quotes > Quotable Quote." *Go-odreads.com*, www.goodreads.com/quotes/206421-men-give-me-credit-for-some-genius-all-the-genius.

[323] Davis, Gary A. "Techniques for Creative Thinking: Yes, They Work." *R & D Innovator*, vol. 1, no. 2, Sept. 1992. Reprinted by Win-stonbrill.com, 2006. www.winstonbrill.com/bril001/html/article_in-dex/articles/1-50/article6_body.html.

324 "Sunni Robertson on how a kingfisher inspired a bullet train." *EarthSky.org*, 29 June 2012, earthsky.org/earth/sunni-robertson-on-how-a-kingfisher-inspired-a-bullet-train.

325 Koba, Susan et al. *Mixing It Up: Integrated, Interdisciplinary, Intriguing Science in the Elementary Classroom.* NSTA Press, 2009.

326 Bellis, Mary. "Who Invented Velcro?" Last updated 1 Apr. 2016, inventors.about.com/library/weekly/aa091297.htm.

327 "Food for Space Flight." *NASA.gov*, 26 Feb. 2004, www.nasa.gov/audience/forstudents/postsecondary/features/F_Food_for_Space_Flight.html.

328 Freeman, R.B. and P.J. Gautrey. "Darwin's *Questions on The Breeding of Animals*, with a note on *Queries about Expression*." Journal of the society for the bibliography of natural history, vol. 5, no. 3, 1969, pp. 220-225, darwin-online.org.uk/converted/pdf/1969_Questions_F1923.pdf.

329 Gassmann, Oliver and Marco Zeschky. "Opening up the Solution Space: The Role of Analogical Thinking for Breakthrough Product Innovation." *Creativity and Innovation Management*, vol. 17, no. 2, 2008, pp. 97-106. papers.ssrn.com/sol3/papers.cfm?abstract_id=1130523.

330 Gassmann, Oliver and Marco Zeschky. "Opening up the Solution Space: The Role of Analogical Thinking for Breakthrough Product Innovation." Creativity and Innovation Management, vol. 17, no. 2, 2008, pp. 97-106. papers.ssrn.com/sol3/papers.cfm?abstract_id=1130523.

[331] Gassmann, Oliver and Marco Zeschky. "Opening up the Solution Space: The Role of Analogical Thinking for Breakthrough Product Innovation." Creativity and Innovation Management, vol. 17, no. 2, 2008, pp. 97-106. papers.ssrn.com/sol3/papers.cfm?abstract_id=1130523.

[332] Gordon, William J.J. *Synectics: The Development of Creative Capacity*. New York: Harper and Row, 1961.

[333] Davis, Gary A. "Techniques for Creative Thinking: Yes, They Work." R & D Innovator, vol. 1, no. 2, Sept. 1992. Reprinted by Winstonbrill.com, 2006. www.winstonbrill.com/bril001/html/article_index/articles/1-50/article6_body.html.

[334] Davis, Gary A. "Techniques for Creative Thinking: Yes, They Work." R & D Innovator, vol. 1, no. 2, Sept. 1992. Reprinted by Winstonbrill.com, 2006. www.winstonbrill.com/bril001/html/article_index/articles/1-50/article6_body.html.

[335] Tannenbaum, Rob. "Paul Simon: Graceland." *Rolling Stone*, 21 Jan. 1997. www.rollingstone.com/music/albumreviews/graceland-19970121.

[336] Kot, Greg. "Space Oddities: David Bowie's Hidden Influences." *BBC.com*, 11 Jan. 2016, www.bbc.com/culture/story/20160108-space-oddities-david-bowies-hidden-influences.

Johnson, Emily. "Kabuki and the Art of...David Bowie?" *InsideJapan*, 11 Jan. 2016, www.insidejapantours.com/blog/2016/01/11/kabuki-and-the-art-of-david-bowie.

[337] "Inventor in History: Eli Whitney." *Intellectual Ventures Laboratory*, 4 Dec. 2013, www.intellectualventureslab.com/invent/inventor-in-history-eli-whitney.

338 Strathern, Paul. *The Medici: Godfathers of the Renaissance.* London: Jonathan Cape, 2003.

339 Johansson, Frans. The Medici Effect: What Elephants and Epidemics Can Teach Us About Innovation. Harvard Business Review Press, 2006.

Johansson, Frans. *The Medici Effect: What Elephants and Epidemics Can Teach Us About Innovation.* Harvard Business Review Press, 2006.

340 Johansson, Frans. *The Medici Effect: What Elephants and Epidemics Can Teach Us About Innovation.* Harvard Business Review Press, 2006.

341 Johnson, Steven. *Where Good Ideas Come From: The Natural History of Innovation.* New York: Riverhead, 2010.

342 Hardesty, Larry. "Why innovation thrives in cities." MIT News Office, 4 June 2013. news.mit.edu/2013/why-innovation-thrives-in-cities-0604.

343 Johnson, Steven. *Where Good Ideas Come From: The Natural History of Innovation.* New York: Riverhead, 2010.

344 Kaufman, Scott Barry and Carolyn Gregoire. "How to Cultivate Creativity." *Scientific American Mind*, 1 Jan. 2016, www.scientificamerican.com/article/how-to-cultivate-your-creativity-book-excerpt.

345 DeYoung, Colin. "The neuromodulator of exploration: a unifying theory of the role of dopamine in personality." *Frontiers in Human Neuroscience*, 14 Nov. 2013, www.tc.umn.edu/~cdeyoung/Pubs/DeYoung_2013_dopamine_personality_Frontiers.pdf.

[346] Kaufman, Scott Barry and Carolyn Gregoire. "How to Cultivate Creativity." *Scientific American Mind*, 1 Jan. 2016, www.scientificamerican.com/article/how-to-cultivate-your-creativity-book-excerpt.

[347] Anyaso, Hilary Hurd. "Creative Genius Driven by Distraction." *Northwestern Now*, 3 Mar. 2015, www.northwestern.edu/newscenter/stories/2015/03/creative-genius-driven-by-distraction.html.

[348] Anyaso, Hilary Hurd. "Creative Genius Driven by Distraction." *Northwestern Now*, 3 Mar. 2015, www.northwestern.edu/newscenter/stories/2015/03/creative-genius-driven-by-distraction.html.

[349] Kaufman, Scott Barry and Carolyn Gregoire. "How to Cultivate Creativity." *Scientific American Mind*, 1 Jan. 2016, www.scientificamerican.com/article/how-to-cultivate-your-creativity-book-excerpt.

[350] Pink, Daniel H. *To Sell Is Human: The Surprising Truth about Moving Others*. New York: Riverhead, 2012.

[351] Fryxell, David A. "History Matters: High Rollers." *Family Tree Magazine*, 16 Dec. 2009, www.familytreemagazine.com/Article-Print/History-Matters-High-Rollers.

[352] Davis, Gary A. "Techniques for Creative Thinking: Yes, They Work." *R & D Innovator*, vol. 1, no. 2, Sept. 1992. Reprinted by Winstonbrill.com, 2006. www.winstonbrill.com/bril001/html/article_index/articles/1-50/article6_body.html.

Chapter Seven

[353] "This is Your Brain on Jazz: Researchers Use MRI to Study Spontaneity, Creativity." Johns Hopkins Medicine, 26 Feb. 2008, www.hopkinsmedicine.org/news/media/releases/this_is_your_brain_on_jazz_researchers_use_mri_to_study_spontaneity_creativity.

354 Kotler, Steven. *The Rise of Superman: Decoding the Science of Ultimate Human Performance*, New York: Houghton Mifflin Harcourt Publishing Company, 2014.

355 Csikszentmihalyi, Mihaly. *Flow: The Psychology of Optimal Experience*. New York: Harper & Row, 1990.

356 Csikszentmihalyi, Mihaly. *Flow: The Psychology of Optimal Experience*. New York: Harper & Row, 1990. Pp. 52

357 Csikszentmihalyi, Mihaly. *Flow: The Psychology of Optimal Experience*. New York: Harper & Row, 1990. Pp. 53.

358 Jackson, Phil, and Hugh Delehanty. *Eleven Rings: The Soul of Success*. Penguin Books, 2014.

359 Kotler, Steven. "Live Life at the Limits: How to Hack Your Flow." *Art of Manliness*, 4 Mar. 2014, www.artofmanliness.com/2014/03/04/live-life-at-the-limits-how-to-hack-your-flow.

360 Kotler, Steven. "Live Life at the Limits: How to Hack Your Flow." *Art of Manliness*, 4 Mar. 2014, www.artofmanliness.com/2014/03/04/live-life-at-the-limits-how-to-hack-your-flow.

361 Dietrich, Arne. *How Creativity Happens in the Brain*. Basingstoke: Palgrave Macmillan, 2015.

362 Dietrich, Arne. Personal interview. 22 Jul. 2016.

363 Dietrich, Arne. *How Creativity Happens in the Brain*. Basingstoke: Palgrave Macmillan, 2015. Pp. 164.

364 Dietrich, Arne. Personal interview. 22 Jul. 2016.

[365] Williams, Neil. "Zlin wing Structural Failure Report." *The British Aerobatic Association*. Reprinted from *Flight International*, 18 Jun. 1970. historic.aerobatics.org.uk/repeats/zlin_wing_failure.htm.

[366] Csikszentmihalyi, Mihaly. *Flow: The Psychology of Optimal Experience*. New York: Harper & Row, 1990. Pp. 66-67.

[367] Csikszentmihalyi, Mihaly. *Finding Flow: The Psychology of Engagement with Everyday Life*. New York: Basic, 1997.

[368] Watts, Alan. *The Essence of Alan Watts*. Celestial Arts, 1977. Pp. 146.

[369] Csikszentmihalyi, Mihaly. *Finding Flow: The Psychology of Engagement with Everyday Life*. New York: Basic, 1997. Pp. 130.

[370] Csikszentmihalyi, Mihaly. *Finding Flow: The Psychology of Engagement with Everyday Life*. New York: Basic, 1997. Pp. 127.

[371] McLaughlin, Katie. "Michael Phelps: 'I consider myself normal'." *CNN.com*, 1 Aug. 2012. www.cnn.com/2012/07/30/us/michael-phelps-on-pmt.

[372] Csikszentmihalyi, Mihaly. *Flow: The Psychology of Optimal Experience*. New York: Harper & Row, 1990.

[373] Csikszentmihalyi, Mihaly and Isabella Csikszentmihalyi. *Optimal experience: Psychological studies of flow in consciousness*. New York: Cambridge University Press, 1988.

Csikszentmihalyi, Mihaly and R Larson. "Validity and reliability of the experience sampling method." Journal of Nervous and Mental Disease, vol. 175, no. 9, 1987, pp. 526-536. www.ncbi.nlm.nih.gov/pubmed/3655778.

Anderson, Lynn. "Use of Experience Sampling Method to Understand the Wilderness Experience." *United States Department of Agriculture Forest Service*, www.nrs.fs.fed.us/pubs/gtr/gtr_ne289/gtr_ne289_092.pdf.

[374] Csikszentmihalyi, Mihaly. *Flow: The Psychology of Optimal Experience*. New York: Harper & Row, 1990.

[375] "American Time Use Survey – 2015 Results." *Bureau of Labor and Statistics*, 24 Jun. 2016. www.bls.gov/news.release/atus.nr0.htm.

[376] Kubey, Robert and Mihaly Csikszentmihalyi. *Television and the quality of life: How viewing shapes everyday experience*. Hillsdale, NJ: Lawrence Earlbaum, 1990.

[377] Csikszentmihalyi, Mihaly. *Finding Flow: The Psychology of Engagement with Everyday Life*. New York: Basic, 1997.

[378] Kotler, Steven. *The Rise of Superman: Decoding the Science of Ultimate Human Performance*, New York: Houghton Mifflin Harcourt Publishing Company, 2014.

[379] McCurry, Justin. "Ancient art of pearl diving breathes its last." The Guardian, 24 Aug. 2006, www.theguardian.com/world/2006/aug/24/japan.justinmccurry.

[380] Cruikshank, Mandy Rae. Personal Interview. 28 Jan. 2016.

[381] Germer, K. Christopher et al, editors. *Mindfulness and Psychotherapy, Second Edition*. Guilford Press, Jul. 2013.

[382] "Complementary, Alternative, or Integrative Health: What's In a Name?" *National Center for Complementary and Integrative Health*, Jun. 2016, nccih.nih.gov/health/integrative-health.

[383] Hurley, Dan. "Breathing In Vs. Spacing Out." *New York Times Magazine*, 14 Jan. 2014, www.nytimes.com/2014/01/19/magazine/breathing-in-vs-spacing-out.html.

[384] "Meditation improves circulation." *Brainwave Research Institute*, 2016. www.brainwave-research-institute.com/meditation-improves-circulation.html.

[385] Ricard, Matthieu et al. "Mind of the Meditator." *Scientific American*, vol. 311, no. 5, 2014, pp. 38-45.

[386] Hasencamp, Wendy. "How to Focus a Wandering Mind." *Greater Good*, 17 Jul. 2013, greatergood.berkeley.edu/article/item/how_to_focus_a_wandering_mind.

[387] Csikszentmihalyi, Mihaly. *Finding Flow: The Psychology of Engagement with Everyday Life*. New York: Basic, 1997.

[388] Ricard, Matthieu et al. "Mind of the Meditator." *Scientific American*, vol. 311, no. 5, 2014, pp. 38-45.

[389] Ricard, Matthieu et al. "Mind of the Meditator." *Scientific American*, vol. 311, no. 5, 2014, pp. 38-45.

[390] Bradt, Steve. "Wandering mind not a happy mind." *Harvard Gazette*, 11 Nov. 2010, news.harvard.edu/gazette/story/2010/11/wandering-mind-not-a- happymind.

Brefczynski-Lewis, J. et al. "Neural Correlates Of Attentional Expertise In Long-term Meditation Practitioners." *Proceedings of the National Academy of Sciences*, vol. 104, no. 27, 2007, pp. 11483-11488, doi:10.1073/pnas.0606552104.

Hölzel, B. et al. "Mindfulness practice leads to increases in regional brain gray matter density." *Psychiatry Research: Neuroimaging*, vol. 191, no. 1, 2012, pp. 36-43, doi: 10.1016/j.pscychresns.2010.08.006.

Tang, Y. et al. "Short-term meditation induces white matter changes in the anterior cingulate." Proceedings of the National Academy of Sciences, vol. 107, no. 35, 2010, pp. 15649-15652, doi:10.1073/pnas.1011043107.

[391] McGonigal, Kelly. *The Willpower Instinct: How Self-control Works, Why It Matters, and What You Can Do to Get More of It*. New York: Avery, 2012.

[392] Corliss, Julie. "Mindfulness meditation helps fight insomnia, improves sleep." *Harvard Health Blog*, 18 Feb. 2015, www.health.harvard.edu/blog/mindfulness-meditation-helps-fight-insomnia-improves-sleep-20150.

[393] Suzuki, Wendy. "An Exercise-Meditation Smackdown." *Psychology Today*, 1 Jun. 2013. www.psychologytoday.com/blog/brain-awakenings/201306/exercise-meditation-smackdown.

[394] Gelb, Michael. *How to Think like Leonardo Da Vinci: Seven Steps to Genius Every Day*. New York, NY: Dell Pub., 2000.

[395] Gelb, Michael. *How to Think like Leonardo Da Vinci: Seven Steps to Genius Every Day*. New York, NY: Dell Pub., 2000.

[396] Heydenreich, Ludwig Henreich. "Leonardo Da Vinci." *Encyclopedia Britannica*, last updated 17 Aug. 2016, www.britannica.com/biography/Leonardo-da-Vinci.

[397] Brooks, John. "9 Sketching Exercises Leonardo Da Vinci Practiced To Achieve Artistic Mastery." *Comfort Pit*, 4 May 2014, comfortpit.com/drawing-exercises-leonardo-da-vinci.

[398] "Leonardo Da Vinci's Last Supper." ItalianRenaissance.org, www.italianrenaissance.org/a-closer-look-leonardo-da-vincis-last-supper.

[399] Gelb, Michael. *How to Think like Leonardo Da Vinci: Seven Steps to Genius Every Day*. New York, NY: Dell Pub., 2000.

[400] "Da Vinci—The Genius." *Museum of Science, Boston*. legacy. mos.org/leonardo/bio.html.

[401] Stone, Madeline. "Look inside the rare Leonardo da Vinci notebook that Bill Gates paid more than $30 million for." *Business Insider*, 13 Jul. 2015, www.businessinsider.com/look-inside-the-co-dex-leicester-which-bill-gates-bought-for-30-million-2015-7.

[402] Switek, Brian. "Leonardo da Vinci - Paleontology Pioneer." *Smithsonian.com*, 11 Jun. 2010, www.smithsonianmag.com/sci-ence-nature/leonardo-da-vinci-paleontology-pioneer-1-73326275.

[403] "Painting in the Style of Old Masters: Sfumato and Chiar-oscuro." About.home, last updated 26 Jul. 2016, painting.about.com/od/oldmastertechniques/a/sfmuato_chiaros.htm;

[404] Gelb, Michael. *How to Think like Leonardo Da Vinci: Seven Steps to Genius Every Day*. New York, NY: Dell Pub., 2000.

[405] "Leonardo Da Vinci." *Biography.com*, last updated 17 Nov. 2015, http://www.biography.com/people/leonardo-da-vin-ci-40396?page=11#synopsis.

[406] NPR staff. "Da Vinci's String Organ Must Be Heard To Be Believed." *NPR Weekend Edition*, 1 Dec. 2013, www.npr.org/sections/deceptivecadence/2013/12/01/247543086/da-vinci-s-string-organ-must-be-heard-to-be-believed.

[407] "Q: How long did it take to paint the lips on the "Mona Lisa"?" *Reference.com,* www.reference.com/art-literature/long-did-paint-lips-mona-lisa-fbe51201498f5a3#.

[408] Powers, Ron. *Mark Twain, Ron Powers.* Simon & Schuster, 2008. pp. 469.

[409] Csikszentmihalyi, Mihaly. Interview by John Geirland. "Go With The Flow." *Wired,* 1 Sept. 1996, www.wired.com/1996/09/czik.

About the Authors

Robert G. Best is Senior Cognitive Strategy Advisor for the Elkay Corporation. Robb's life experience is truly an American gumbo: railroad worker, land surveyor, musician, woodworker, public school teacher, award winning sales/designer and consultant.

Robb also ran a nationally recognized company. His fascination with cognitive science led to the development of a selling and marketing system based on cutting edge brain research. Robb appears throughout the United States and abroad as a keynote speaker. His workshops and seminars are booked out a year in advance and his science blog, *Mindframewithrobb,* is regularly read in 14 countries.

J.M. Best is a freelance writer and editor, with clients in marketing and education. She also writes, directs, and executive produces the sci-fi podcast *The Strange Case of Starship Iris.*

Visit us at www.bestmindframe.com

CPSIA information can be obtained
at www.ICGtesting.com
Printed in the USA
FFOW05n2229150817